Ship To:

AMZ APTP ORDER
800 AVONDALE AVE
COLUMBUS, OH 43212-3473

AF894987

Order ID: 112-8760991-4843440

Thank you for buying from QualityCounts4U2 on Amazon Marketplace.

Shipping Address:	Order Date:	Thu, Oct 22, 2026
AMZ APTP ORDER	Shipping Service:	Standard
800 AVONDALE AVE	Buyer Name:	Peter
COLUMBUS, OH 43212-3473	Seller Name:	QualityCounts4U2

Quantity	Product Details	Unit price	Order Totals
1	**M5 Water Utility Management, Third Edition [Paperback] [2018] AWWA** **SKU:** AWWA M5 Third Edition **ASIN:** 162576247X **Condition:** Used - Like New **Order Item ID:** 173100097385201 **Condition note:** This USED 8 1/2" x 11" coverbind bound protective cover pdf is in like new condition. All pages are clean, white, and free of underlines or margin notes. The manual is bound with a transparent plastic front and black back. The bound protective cover provides protection from scratches, wear, and provides longevity of the document.	$104.50	**Item subtotal** $104.50 **Item total** **$104.50**

Grand total: $104.50

Returning your item:

Go to "Your Account" on Amazon.com, click "Your Orders" and then click the "seller profile" link for this order to get information about the return and refund policies that apply.

Visit https://www.amazon.com/returns to print a return shipping label. Please have your order ID ready.

Thanks for buying on Amazon Marketplace. To provide feedback for the seller please visit www.amazon.com/feedback. To contact the seller, go to Your Orders in Your Account. Click the seller's name under the appropriate product. Then, in the "Further Information" section, click "Contact the Seller."

Manual of Water Supply Practices

M5

Water Utility Management

Third Edition

American Water Works Association

Manual of Water Supply Practices

M5

Water Utility Management

Third Edition

American Water Works Association

Manual of Water Supply Practices—M5

Water Utility Management, 2nd edition

Project Manager: Melissa Valentine
Cover Art: Melanie Yamamoto
Production: PerfecType, Nashville, TN

If you find errors in this manual, please email books@awwa.org. Possible errata will be posted at www.awwa.org/resources-tools/resource.development.groups/manuals-program.aspx.

Library of Congress Cataloging-in-Publication Data

Names: Ginley, James F., author. | Klein, Matthew (Water utility expert), author. | American Water Works Association, author.
Title: Water utility management / by Jim Ginley and Matthew Klein.
Other titles: M5, water utility management
Description: Third edition. | Denver, CO : American Water Works Association, [2018] | Series: AWWA manual ; M5 | Previous editions by American Water Works Association. | Includes bibliographical references and index.
Identifiers: LCCN 2017049003 | ISBN 9781625762474 (alk. paper)
Subjects: LCSH: Water utilities--Management--Handbooks, manuals, etc. | Waterworks--Management--Handbooks, manuals, etc.
Classification: LCC HD4456 .G56 2018 | DDC 363.6/1068--dc23
LC record available at https://lccn.loc.gov/2017049003

ISBN: 978-1-62576-247-4
eISBN: 978-1-61300-434-0

Printed in the United States of America
American Water Works Association
6666 West Quincy Avenue
Denver, CO 80235-3098
awwa.org

Printed on recycled paper

Contents

Figures

This page intentionally blank.

Tables

This page intentionally blank.

Acknowledgments

This edition of the M5 Manual, *Water Utility Management*, was developed by Jim Ginley, past chair of the Strategic Management Practices Committee. His career of nearly 30 years in the water sector as a not-for-profit association professional, utility management consultant, regulatory compliance specialist, and active volunteer contributed to the organization of many chapters. Matthew Klein, chair of Strategic Management Practices Committee and president of North Carolina and Tennessee Utilities Inc., provided new and creative solutions for the updating of M5.

The Strategic Management Practices Committee gives special thanks and recognition to the following individuals who contributed as primary authors and significant content contributors: Billie Blair, Seth Garrison, Jim Ginley, David Hanes, Stuart Karasik, Matthew Klein, Jay Madigan, David McGimpsey, Kevin Morley, Peter Nardi, Eric Rothstein, Karen Snyder, Patricia Tennyson, and Pieter Van Ry.

Additionally, the Strategic Management Practices Committee acknowledges with gratitude the peer review, suggestion of content, and editorial, administrative, and other contributions of the following volunteers: Lori Armstrong, Colleen Arnold, Alvin Bautista, Robert Boswell, Fred Bloetscher, Mike Caston, Julius Ciaccia, Catherine Curtis, Maria DeChellis, Tom DeLaura, Dale Jutila, Lee Anne Jones, Elizabeth Kelly, Prabha Kumar, Venus Larson, Manshi Low, Brian Mantz, Sharon Peters, Connie Schreppel, Doug Spiers, Mark Stanley, and Orville Watson. Finally, a special note of thanks to Marta Madigan for her graphic artistry.

This page intentionally blank.

Acronyms and Abbreviations

AMR/AMI	automated meter reading/advanced metering infrastructure
AMWA	Association of Metro Water Agencies
APD	alternative project delivery
APQC	American Productivity and Quality Center
ASP	Association for Strategic Planning
AWWA	American Water Works Association
BI	business intelligence
CAA	Clean Air Act
CCR	consumer confidence report (eCCR = electronic consumer confidence report)
CERCLA	Comprehensive Environmental Response, Compensation, and Liability Act
CIP	capital improvement plan
CIS	customer information system
CMMS/EAM	computerized maintenance management system/enterprise asset management
CPM	capital program management
CSO	combined sewer overflow
CSR	customer service representative
CUPSS	Check Up Program for Small Systems
CWA	Clean Water Act
CWSRF	Clean Water State Revolving Fund
DCS	distributed control system
DHS	Department of Homeland Security
DWSRF	Drinking Water State Revolving Fund
EPCRA	Emergency Planning and Community Right-to-Know Act
ERP	enterprise resource planning
ETA	Employment and Training Administration
EUM	effective utility management
FAQ	frequently asked question
FEMA	Federal Emergency Management Agency
FERC	Federal Energy Regulatory Commission
FIFRA	Federal Insecticide, Fungicide, and Rodenticide Act
FIS	financial information system
GIS	geographic information system
HMI	human-machine interface
HSE	health, safety, and environment
ICS	Incident Command System
ICS-CERT	Industrial Control Systems Cyber Emergency Response Team
I/I	infiltration and inflow
IoE	Internet of Everything
IoT	Internet of Things
IT	information technology
IWA	International Water Assocation
KM	knowledge management
LIMS	laboratory information management system
LIS	laboratory information system

LOS	levels of service
MCL	maximum contaminant level
NAAQS	National Ambient Air Quality Standards
NACWA	National Association of Clean Water Agencies
NCEES	National Council of Examiners for Engineering and Surveying
NIMS	National Incident Management System
NIST	National Institute of Standards and Technology
NPDES	National Pollutant Discharge Elimination System
ODMS	operations data management system
PC	personal computer
PLC	programmable logic controller
PM	preventive maintenance
POTW	publicly owned treatment works
RCM	reliability-centered maintenance
RCRA	Resources Conservation and Recovery Act
RFID	radio-frequency identification
ROI	return on investment
RTU	remote terminal unit
SaaS	software-as-a-service
SCADA	supervisory control and data acquisition
SDC	system development charge
SDWA	Safe Water Drinking Act
SIMPLE	Sustainable Infrastructure Management Program Learning Environment
SMART	specific measurable achievable relevant and time based
SSO	sanitary sewer overflow
TSCA	Toxic Substances Control Act
URI	Utility Resilience Index
US BLS	US Bureau of Labor Statistics
US DOL	US Department of Labor
USEPA	US Environment Protection Agency
UST	underground storage tank
VA	vulnerability assessment
WARN	Water and Wastewater Agency Response Network
WaterRF	Water Research Foundation
WEF	Water Environment Federation
WERF	Water Environmental Research Foundation
WIFA	Water Infrastructure Financing Authority

Introduction

THE NEW M5: FACILITATING SUCCESS FOR WATER AND WASTEWATER UTILITY MANAGERS

Can you imagine a city without water and wastewater utilities? Imagine whole towns drinking untreated water. Can you imagine US households without access to functioning sewer systems? The simple truth is that our modern lives are built on the premise that we will have clean water and the ability to dispose of water 24 hours a day, every day of the year. The provision of water services is fundamental to us—a matter of public health.

North American water and wastewater managers are public health providers with unparalleled track records of success. Another simple truth—in the twentieth century alone, the applications of our industry's science, tools, and trades saved more lives than any other single health development. The people who design and manage our water processes are nothing less than heroes and our target readers of the updated M5. The M5 update is here to help water and wastewater utility managers stay heroes in the twenty-first century.

Today, increasing demands, changing technologies, uncertain costs, and aging systems complicate every utility manager's role. Whether a utility is large or small, successful utility management requires more than single-subject expertise. High expectations are placed on established managers to improve their range of functionality. M5 will help experienced managers in North America and elsewhere broaden their perspectives. Newly minted utility managers will find here a means to untangling today's multiple water–wastewater utility issues and to continue taking steps toward a fulfilling career.

A GO-TO REFERENCE FOR WHAT IS IMPORTANT TO YOU

Utility managers need a strategic process for establishing their vision, mission, and plan for achieving their most critical objectives. Chapter 3 is dedicated to utility strategy. Need a primer on performance measurement? See chapter 5. Where do you learn about establishing a cycle (a culture) of plan/do/check/act, and do that with an ethnically and educationally diverse workforce? Turn to chapter 4. Want to survey your employees in accord with the 10 attributes of the effective utility management (EUM) approach? Consider chapter 6. The updated M5 will help locate the best management resources.

This manual delivers approaches to the plans (and planning processes) that require more detail than overarching strategic planning. Managers also need to account for their specific responsibilities, which they may choose to do in a communication plan, a capital plan, a divisional business plan, and master planning of all varieties (see chapter 3). This update is designed as your road map to navigate the array of resources that are already available—in print, in electronic media, at conferences, and with peer utilities.

UPDATING M5

AWWA's Strategic Management Practices Committee was tasked with updating M5. The committee began by studying the contemporary topics of water–wastewater utility managers, including workforce capability; fostering safety culture; managing assets and risk; the tricky balance of public support, scarce revenues, and aging infrastructure; and rapidly changing information management technology.

The committee built on the previous edition of M5, which documented the fundamentals of managing a utility from managing operations and maintenance to customer service, from communication to emergency planning, and from safety regulation to financial management. By focusing on the key contemporary topics of vital interest to water–wastewater managers, the committee has produced a new edition of M5 that goes beyond basic books on water–wastewater utility management. Additionally, the committee has included information on the vast resources that exist in AWWA on these well-established management topics. And it should be noted that books in the general management literature covering business strategy, leadership, and a variety of management techniques also have merit for utility managers. Finally, while the committee's experience and strengths were gained from a North American perspective, utility managers will benefit no matter where they are based.

Since the most recent M5 update, there has been increasing demand for water and wastewater utilities to focus on continuous improvement. Improvement programs may include self-assessment, outside assistance (peer review or consultancy), and benchmarking tools. QualServe was an early example of a continuous improvement system designed specifically for water and wastewater utilities. Involving self-assessment, peer review, and benchmarking, this program provided the foundation for *Effective Utility Management: A Primer for Water and Wastewater Utilities.*

The EUM *Primer* defines attributes common to all water and wastewater utilities, covering a range of desired utility outcomes in the areas of operations, infrastructure, customer satisfaction, community welfare, natural resource stewardship, and financial performance. These attributes have proven useful and concise reference points for utility managers seeking to improve organizationwide performance. They can best be viewed as a continuum of, or a set of building blocks for, management improvement opportunities and are referenced throughout M5.

Chapter 1

Utility Management and Regulation

This chapter is a high-level review of the legal basis of regulation and management approaches. The purpose is to describe fundamental regulatory considerations faced by a water or wastewater utility manager.

Wastewater and drinking water utilities are driven by a mission to ensure public health through water management of all kinds. A manager's planning agenda is frequently dominated by working toward objectives associated with public health; compliance with legal obligations, including federal, state, and local; and environmental, health, and safety requirements. Failure to comply with many of these requirements may result in administrative, civil, and criminal penalties, in addition to unwanted attention from the media.

Utility managers are leaders of multitalented people out of necessity, because no one knows everything about water. Clean or dirty water management requires engineering, biological, chemical, design, financial, organizational, legal, and other subject matter expertise. Variations in regulatory requirements mean utility managers should have ready access to qualified professionals in nearly every field. Legal expertise is an important, even essential collaborating partner to a utility manager.

This manual is one of multiple resources of the American Water Works Association (AWWA) designed specifically for water and wastewater professionals. Importantly, AWWA creates standards addressing the entire industry spectrum of treatment plant operators and managers, scientists, environmentalists, manufacturers, academicians, regulators, and utilities. AWWA's published standards cover hundreds of products and procedures.

LOOKING FOR LAW IN ALL THE RIGHT PLACES

Environmental legal obligations are numerous and complex. These obligations exist at the federal, state, and local levels. The nature of the legal obligations imposed on utilities ranges from broad statutes to complex regulations or policies. The legal obligations address the relationship of the utility's operations to the health and safety of its employees and its impact on the environment. As a result, it is suggested that a leader of a water or wastewater utility have access to a qualified environmental attorney or, as necessary, an environmental regulatory professional. In consultation with a qualified environmental attorney, it is likely that a water or wastewater utility will better address environmental legal obligations and attempt to prevent, detect, or correct any violations alleged by federal, state, or local authorities.

Utility managers must possess a fundamental understanding of the major environmental legal obligations impacting utility operations. Depending on geographic location and size of operations, water and wastewater utilities' legal obligations to the variety of federal, state, or local environmental requirements may vary. These environmental legal obligations certainly include well-known statutory or regulatory provisions of the Safe Drinking Water Act (SDWA), which includes the provisions for National Pollutant Discharge Elimination System (NPDES). The NPDES addresses water pollution by regulating point sources that discharge pollutants to waters of the United States. This is a permit program authorizing state governments to perform many permitting, administrative, and enforcement aspects of the USEPA-administered program.

It should be recognized that many state and/or local environmental authorities have implemented and will enforce provisions either equal to or more stringent than those contained within the Clean Air Act (CAA), the Clean Water Act (CWA), the Resource Conservation and Recovery Act (RCRA), the Comprehensive Environmental Response, Compensation, and Liability Act (CERCLA), the Federal Insecticide, Fungicide, and Rodenticide Act (FIFRA), the Emergency Planning and Community Right-to-Know Act (EPCRA), and the Toxic Substances Control Act (TSCA).

Drinking water and wastewater operation and treatment facilities are part of the utilities sector designated by the US Environmental Protection Agency (USEPA) in the North American Industry Classification System known as NAICS 22. Industries in this sector include water treatment plants, water supply systems, sewer systems, and/or sewage treatment facilities that collect, treat, and dispose of waste.

HISTORY IN YOUR LIFETIME

The Clean Water Act, a crucial law that protects the nation's water from pollution, was passed at a time when much of US water was so contaminated by industrial waste and other pollutants that it was unfit for public use. By setting ambitious goals for the cleanup of contaminated waters, the Clean Water Act led to dramatic improvements in water quality and serious reductions in industrial pollution. Adoption of the Clean Water Act followed several high-profile disasters like Ohio's Cuyahoga River catching on fire because of the contaminants in the water. The Clean Water Act strengthened the statutory framework and required mandatory pollution controls and meaningful enforcement mechanisms.

The Clean Water Act set a new national goal "to restore and maintain the chemical, physical, and biological integrity of the Nation's waters." It sought to ensure all waters be "fishable and swimmable" and provided the USEPA and states with the authority to set and implement the standards necessary to achieve these goals. Generally, states set water quality targets and standards that define specific cleanup measures or limit the amount of pollution that can be discharged into bodies of water; USEPA then reviews and approves these targets and standards. The Clean Water Act also established many different programs aimed at protecting wetlands, coastal waters, estuaries, and large ecosystems.

The year 2014 marked the fortieth anniversary of the SDWA. Utility managers know this as the law USEPA uses to establish and enforce national primary drinking water standards. These standards are known as the National Primary Drinking Water Regulations and set enforceable maximum contaminant levels (MCLs) for particular contaminants in drinking water or require ways to treat water to remove contaminants. Each standard also includes requirements for water systems to test for contaminants in the water to make sure standards are achieved. USEPA has established MCLs or specific treatment techniques for more than 80 contaminants and has also developed (unenforceable) secondary drinking water standards to address the protection of the public welfare (e.g., taste-and-odor issues).

The SDWA was amended in 1986 and again in 1996, expanding both the list of contaminants to be monitored and the scope of activities required to protect public drinking water and its sources (including lakes, rivers, reservoirs, springs, and groundwater wells). The original SDWA focused primarily on treatment as the means of providing safe drinking water at the tap. The amendments enacted in 1996 greatly enhance the concept of protecting drinking water from source to tap. To accomplish this, the amendments call for additional attention to be paid to protecting source water, training certified operators, establishing funds for water system improvements, and using public information to ensure safe drinking water. For an excellent overview, see the USEPA fact sheet titled "Understanding the Safe Drinking Water Act" (USEPA 2004).

The SDWA applies to every public water system in the United States and its territories (e.g., Puerto Rico, the US Virgin Islands, and Guam) and provides the statutory and regulatory framework to ensure that the requirements are both completed and enforced. In most cases, oversight and enforcement of water systems are the responsibility of the respective state drinking water programs. To formally assume this responsibility, states can apply to the USEPA for *primacy,* which gives them the authority to implement the SDWA requirements within their jurisdictions. To do this, states must be able to show that they will adopt standards at least as stringent as USEPA's and make sure water systems meet these standards. The Navajo Nation is the only tribal entity with SDWA primacy.

As of 2017, only Wyoming and the District of Columbia have not received primacy. States, or the USEPA acting as a primacy agent, make sure water systems test for contaminants, review plans for water system improvements, conduct on-site inspections and sanitary surveys, provide training and technical assistance, and take action against water systems not meeting standards.

The State Perspective Matters

Given the role of the state primacy agencies, it is essential for utility leaders to know all state-based requirements *and* to be familiar with the agency that is responsible for making sure a utility is in compliance. In most cases, the relationship between the water–wastewater utility and the state agency is well established. You should ensure managers and key operations staff get to know the state agency and the staff assigned to a given utility. Make sure you are aware of all required reports and communications and establish a regular dialogue with the key members of your state agency.

Other Key Federal Statutes

Another key federal statute usually enforced by states is the CWA of 1972. The purpose of this law is to address the pollution of water in the United States. Among other things, the CWA contains provisions regulating municipal wastewater treatment systems. Along the same lines as the SDWA, the USEPA has established and enforces requirements broadly prohibiting the discharge of pollutants into waters of the United States without a permit, referred to as an NPDES permit. For example, the discharge of treatment filter backwash residuals into an adjacent river either without an NPDES permit or in violation of a condition of a valid NPDES permit may likely result in an enforcement action by an appropriate authority. Notably, the CWA also addresses (1) the discharge or placement of dredged or fill material into waters of the United States (i.e., wetlands issues); and (2) the prevention of and responses to spills of oil and hazardous substances.

RCRA (1976) is another key federal statute. In general, RCRA is characterized as the federal statute providing "cradle-to-grave" management and tracking of hazardous waste (e.g., spent solvents). The USEPA has established enforceable standards for the identification, management, transportation, and disposal of hazardous waste. Other portions of RCRA address (1) used oil management and (2) underground storage tank (UST) systems.

Importantly, the CWA prohibits anybody from discharging pollutants through a point source into a water of the United States unless they have an NPDES permit. The permit will contain limits on what can be discharged, what can be monitored, what are the reporting requirements, and other provisions to ensure that the discharge does not hurt water quality or people's health. In essence, the permit translates general requirements of the CWA into specific provisions tailored to the operations of each person discharging pollutants.

Another significant federal statute is the CAA (1970), which addresses air emissions from both stationary and mobile sources. The key component of the CAA is the National Ambient Air Quality Standards (NAAQS). Through the NAAQS program, USEPA regulates six ambient air pollutants determined to endanger public health and welfare (e.g., particulate matter). The CAA also addresses hazardous air pollutants and risk management planning. Drinking water utilities may possess certain operations requiring either CAA construction and/or operation permits (e.g., diesel generators) or risk management plans (e.g., chlorine cylinders).

Canada and Mexico

In Canada, the federal government has jurisdiction related to fisheries, navigation, federal lands, and international relations, including responsibilities related to the management of boundary waters shared with the United States and relations with the International Joint Commission. It also has significant responsibilities for agriculture, health, and the environment; plays a significant role supporting aquatic research and technology; and ensures national policies and standards are in place on environmental and health-related issues. Within the Canadian government, more than 20 departments and agencies have unique responsibilities for fresh water.

Water supply and sanitation in Mexico are characterized by achievements such as a significant increase in access to piped water supply in urban areas (88 percent to 93 percent) as well as in rural areas (50 percent to 74 percent) between 1990 and 2010. Additionally, a strong nationwide increase in access to improved sanitation (64 percent to 85 percent) was observed in the same period. Other achievements include the existence of a functioning national system to finance water and sanitation infrastructure with a National Water Commission as its apex institution, and the existence of a few well-performing utilities such as Aguas y Drenaje de Monterrey. In addition to ongoing investments to expand access, the government has embarked on a large investment program to improve wastewater treatment.

REFERENCES

USEPA. 2004. "Understanding the Safe Drinking Water Act." EPA 816-F-04-030, at http://water.epa.gov/lawsregs/rulesregs/sdwa/index.cfm.

Chapter 2

Utility Management Overview

This chapter describes the nontechnical role of management and surveys management approaches. The purpose is to describe parameters of how a water–wastewater utility manager conducts the business of protecting public health and the consequent economic viability of the community. Included are the basics of developing a strategic plan, as well as an example of a simple strategic planning framework based on the effective utility management (EUM) "attributes of effectively managed utilities." Here also you will find suggestions on how to involve staff and stakeholders by tracking and reporting progress and success and by keeping your plan both current and continuously effective by revisiting and revising the plan on a regular basis.

From a technical and operational perspective, water–wastewater utility sector managers are stewards of water throughout its entire cycle—from the source (well, river, lake, reservoir), out to the community and back, and then back to the world again (river, lake, ocean, injection well, landfill, or farmer's field). In sector terminology, this encompasses supply, treatment, distribution, collection, treatment, disposal, and even reuse. This is the physical structure, or better, the infrastructure that encompasses how utilities deliver water, sewer, storm, and environmental services to their communities. In addition, adequate fire flow for residential, commercial, and industrial customers is also important.

More specifically, water utility managers are responsible for the daily operation of one of the most important assets in the community—the infrastructure and processes that harness, treat, and distribute water to the community. Inherent in this responsibility is the fundamental task of providing an adequate supply of potable drinking water and, in doing so, protecting the public health of those who consume that water.

WHAT EVERYONE FACES TODAY

The most prevalent management issues or, in some cases, trends facing water and wastewater utilities are addressed in cycles of strategic planning. Strategic planning initiatives, such as workforce management, influence the development of a utility's use of information, the technology required to manage its data, and how it plans for the care of its assets. In turn, strategic planning requires utilities to define future resources. The ability to communicate the value of a utility's investment in water infrastructure and services is another critical step in strategic planning, shown in Figure 2-1.

PLANNING AND STRATEGY

Organizations and businesses are successful for a variety of reasons: very talented people; great business concept; clever advertising campaigns; a savvy and aggressive sales force; or perhaps fundamentally sound and optimized manufacturing and distribution practices. Whatever the secret to success might be, one thing that most successful organizations and businesses have in common is the development and implementation of a sound business strategy. At the core of every good strategy is sound planning.

Water utility managers are familiar with plans—from routine work plans for the daily work of crews to long-term facility master plans that identify the needs of the organization for decades. Plans are common in many aspects of utility operations, both on the technical side and the management side of the organization. Much like other businesses, many utilities have developed a strategic plan to guide the organization.

DEALING WITH THE REALITIES OF THE CHANGING WORKFORCE

Making sure you have the right people with the right skill sets is important for any organization. In particular, for organizations like utilities that may have a wide variety of highly technical processes and practices, the right skills are imperative for efficient and effective operations. As a utility manager, one of your key roles and most pressing challenges is finding, training, and keeping the right people for the right jobs that contribute to a successfully run utility.

Chapter 4 covers a spectrum of workforce issues—both in the general North American population and specific to the water utility sector. These issues range from how to address the changing demographics of the workplace to how to find, train, and develop the next generation of utility employees. The issues include not only how to capture the vast amount of institutional knowledge and retain it, but also how to use it to train and build a talented, adaptable, and continually learning workforce.

The chapter also focuses on operator issues and some of the tricky realities of replacing a departing generation with highly established skills and a highly evolved sense of role and purpose with a group of people with different backgrounds, different expectations, and different views about the workplace.

MANAGING INFORMATION AND ASSETS

Information management and data management are urgent skills either for you to develop within your organization or for you to have immediate access to through contractual arrangements. From smartphones and iPads in our personal lives to any number of databases, software applications, and data-driven processes in the treatment plant, business office, or customer service center, we use, process, and manage data and information. Data and information are an essential part of your role as a utility manager. You rely on accurate data and information to make decisions, direct work, and make reports both up and down in the organization.

Chapter 5 covers the demanding challenge of managing data and information management in a water utility setting. The first portion of the chapter outlines some of the

Figure 2-1 Strategic management of information

basics of how data flows throughout a utility organization. The second part details managing data across the many business processes found in a modern utility. The last part highlights some of the major trends in information management, including "Big Data," cloud technology and security, mapping interfaces, social media, and business intelligence. The chapter concludes with a focused consideration of utility security issues anchored by a proactive approach.

Continuing the theme of managing important information, chapter 6 covers some of the key aspects of asset management. This chapter begins with the basics, such as taking an inventory of assets, establishing the levels of service you will provide for your assets, and then conducting the risk/condition assessment to help set your priorities. After the basics are covered, the chapter then details how these practices and approaches link with capital planning, information management, and routine operations and maintenance.

VALUE OF WATER IN LIGHT OF VARYING ABUNDANCE AND SCARCITY

Perhaps one of the most pressing issues for water utility managers is trying to explain the value of water to customers, ratepayers, politicians, and the public in general. In some regions of the United States, like California, Texas, Nevada, and other parts of the arid Southwest, utilities and the communities they serve face significant and ongoing drought conditions. In other parts of the United States and Canada, communities have experienced record-breaking storm events—both in terms of quantity of precipitation and ferocity of the storms. And then there is the mounting cost of replacing aging infrastructure—especially for those communities that have old distribution systems carrying the water out and are also addressing consent decrees to fix inadequately designed sanitary and combined sewer systems. One community does not have enough water and the next has too much; another community is mandated to increase rates to fix a problem, and another is facing reduced revenues from either necessary conservative consumption or lack of use because of record rainfall.

Chapter 7 discusses this tricky interaction between the operational realities and demands on a utility, the finances needed to support these operations, and the support that is necessary to garner from ratepayers and governance to successfully establish a viable and sustainable financial situation for the utility.

Chapter 8 picks up where chapter 7 leaves off and carries the conversation about communications further, explaining the importance of effective communications and outreach as important tools in utility management. This includes a discussion about developing and implementing communication plans and strategies, as well as ensuring that you have the right people to develop and provide the right message.

STATUTORY, REGULATORY, AND POLICY ISSUES

All utilities are responsible for complying with the applicable federal, state, and local statutes and regulations that govern their operations. These statutes and regulations cover a full spectrum from how water is obtained and to what degree the water is purified to make it safe for the public to how compliance is proven and reported. Add to these issues the need to provide adequate fire flow for public safety, the need to comply with accounting and business rules, and a long list of rules and requirements that apply to all organizations regardless of the service they provide.

As a result, all utility managers need to understand both the breadth and depth of the statutory and regulatory requirements that fit with their role and also to a large degree those that apply to the entire organization.

Accordingly, chapter 9 provides an overview of environmental health and safety regulations—again with the focus on describing the broad requirements on the statutory and regulatory front, along with some discussion of the importance of communication with regulatory agencies, governing bodies, and the public about how the utility is meeting all of the many requirements.

EVERYDAY MANAGEMENT EXCELLENCE

Managing the organization requires attention to these nuts-and-bolts topics essential to every utility:

- **Customer service**
  - Expectations of internal and external stakeholders
  - Managing the customer service workforce
  - Defining customer service responsibilities
  - Business process management side of customer accounts
- **Operations and maintenance (O&M)**
  - Protection of source water
  - Water treatment plant operations and maintenance
  - Distribution system operations and maintenance

- **Health and safety**
  - Integrating safety into operations and maintenance
  - Injury prevention strategies
  - Regulatory compliance
  - Establishing a safety program for staff and contractors
- **Emergency management**
  - Risk and resilience management
  - Water/wastewater agency response network (WARN)
  - Security practices
  - Emergency preparedness practices
  - Utility resilience index (URI)
- **Financial management**
  - Fiduciary guidelines, regulations, and constraints
  - Everyday business operations of the utility
  - Planning and financial management
  - Running the utility "like a business" and as a transparent public agency
  - Tracking and reporting performance, both internally and externally

As utility managers, you are "called to be responsible stewards of the natural, physical, financial, and human resources entrusted to your care." Much of that responsibility relies on the technical and operational skills your position demands. Many AWWA manuals, guidance documents, publications, periodicals, and training programs address these same management topics. Lists of these references have found a place in M5 because they are so fundamental to the everyday operation of a water–wastewater utility that you need to plan to employ them directly or indirectly every day on the job. The following general reference section lists some important management resources, including general management texts that are useful.

GENERAL REFERENCES

Bloetscher, F. 2011. *Utility Management for Water and Wastewater Operators.* Denver, CO: AWWA. Introduction to water utility management principles and practices for new managers, as well as anyone who wishes to move into utility management.

Covey, S.R. 1991. *Principle-Centered Leadership.* New York: Simon and Schuster. One of Covey's classics that focuses on how to develop and implement a long-term inside-out approach to developing people and organizations.

Kaplan, R.S., and D.P. Norton. 2001. *The Strategy-Focused Organization.* Brighton, MA: Harvard Business Review Press. From the creators of the performance management tool called the *balanced scorecard,* this book introduces a new approach that makes strategy a continuous process that involves both managers and staff at all levels of the organization.

Welch, J., with J.A. Byrne. 2003. *Jack: Straight From the Gut.* New York: Warner Books. A compelling read that is both an autobiographical account of Jack Welch, the iconic leader of General Electric, and a management and leadership book that applies broadly to managers in all sectors.

This page intentionally blank.

Chapter 3

Strategic Planning for Water Sector Utilities

This chapter describes the steps necessary for creating a strategic management approach. The purpose is to provide the water or wastewater utility managers a framework for starting their own strategic process.

WHY DO WE NEED A *STRATEGIC* PLAN?

As water sector agencies, we are accustomed to having plans in our everyday operations. From master plans that lay out our facilities and supplies for decades at a time to detailed plans for information management, asset management, financial management, and so on—all of which are mentioned elsewhere in this text. We are a culture of planners.

When it comes to strategic planning, however, utilities may struggle because strategic planning is a different kind of beast. Based historically in military settings and then more recently in corporate/business environments, strategy-based planning is just that—strategic in nature. At its core, the strategic plan's purpose is to outmaneuver or even overcome an adversary or competitor. While this may serve the military and competitive businesses well, it may not be a great match with public agencies and not-for-profit organizations that exist more to provide necessary services than to outcompete or "win."

That being said, utilities are not off the hook when it comes to strategic planning. Like the military and competitive businesses, all utilities are faced with the reality of a rapidly and ever-changing world—be it a result of regulatory demands, or the challenges of replacing aging infrastructure in an uncertain economy, or juggling the myriad of issues related to workforce, customers, and stakeholders.

Furthermore, utilities as organizations are tasked to provide an essential set of services (including public water supply, fire protection, sanitation, and stormwater drainage) *and* in the process, to protect both public health and the environment. Combined, these make the communities we serve a nice place to live, work, raise a family, run a business, and so on.

This is a tall order to deliver in and of itself. And, if that were not enough, the world in which you manage your utility demands that you be effective and efficient with the resources that you have been entrusted to manage. As a result, you are being held more and more accountable as an organization and as a leader, and you need a game plan to help you chart, navigate, report, and rechart your course of action.

Bottom line, you need a strategic plan to help you manage the responsibilities and changes that you are encountering daily, monthly, seasonally, and long term. That plan needs to be a solidly built strategy to navigate this sea of responsibilities and never-ending changes. Having that strategy documented is essential to your utility's success as an organization and your success as a leader or manager of that organization. The plan cannot be one that is written in a day and parked on a shelf while you continue to do things the "same ol', same ol'" way as before.

NOT JUST A WRITTEN PLAN—A PROCESS, TOO

Effective strategic plans do not come in a box nor can they be downloaded from AWWA or Amazon. Like any good plan, they need to be built and implemented and ultimately reviewed and revised. This effort takes an overall process that must include some kind of (1) foundation of understanding, (2) vision, and (3) a road map to the future, as shown in Figure 3-1.

Before jumping into the details of how to build and implement an effective strategic plan, it is important to share a few more thoughts about the planning process and its fundamental purpose, as well as some basic vocabulary that will be used frequently from this point forward.

Vocabulary

First, let us review some of the vocabulary and terminology of strategic planning. Whether you consult Wikipedia or Harvard Business School, the fundamental components of a strategic plan are vision, mission, values, and strategy.

A *vision* or *vision statement* is an expression of what the organization strives to become or, in other words, what they want to look like, be recognized for, or be known as in the future. By comparison, the *mission* or *mission statement* is an explanation of the fundamental purpose of the organization. That is, why does it exist and what purpose or role does

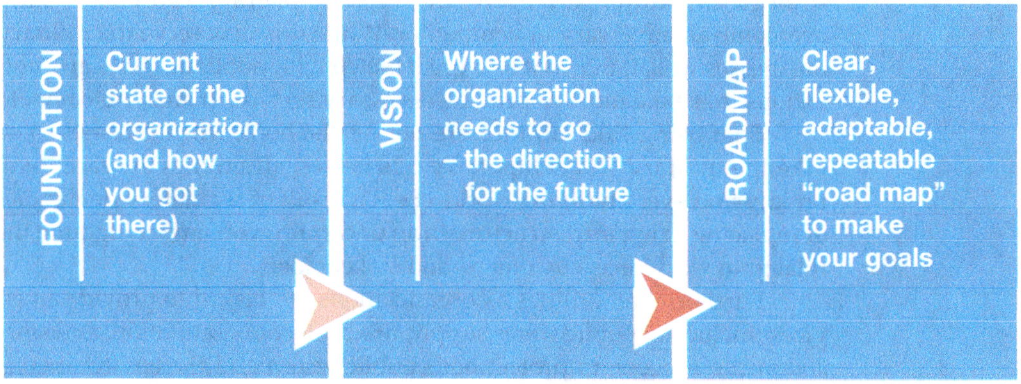

Figure 3-1 The three steps to a plan

the organization serve in the community? The mission statement explains who we are and what we do.

Values or value statements are characteristics, traits, or attributes of an organization that speak broadly and/or specifically to its culture, reputation, and even attitude or disposition. Values can describe how an employee embraces an ethic like "customer service oriented" or what the staff is known for professionally, such as knowledgeable, responsible, or responsive.

Strategy speaks to a collection of activities and words/terms to describe them. These vary from organization to organization and plan to plan. Terms such as *goals* and *objectives*, for example, can be used interchangeably or in cascading fashion. Similarly, the terms *initiative* or *tactics* or *action items* can also be used interchangeably, as well as *metrics* and *measures*. Regardless of the version or variations selected, the bottom line is that the core strategy of the plan is developed around identifying what the organization wants, needs, or intends to do in a defined period of time that both fulfills its established mission and works toward the achievement of its established vision, all the while keeping with the values and the culture of the organization. Whether you call something really important a "goal" or an "objective" is not critical; what is critical is that what you identify is strategically important, has purpose and value relative to the mission and vision of the organization, and can be done responsibly (and perhaps affordably) with the resources you have allocated to the effort.

For the purposes of this text, we will use the following terminology for the components of a "typical" strategic plan:

- **Mission, vision, and values**—as defined previously.
- **Goals**—the most fundamental and important pieces of the strategic plan. These are the anchors of the plan—the handful of (plus or minus) key things that must be achieved in order to fulfill the mission and achieve the vision. These are intentionally high level and even lofty in nature. A good example of a well-written goal might be, "to produce drinking water of the highest quality." This only answers the question, "what are we going to do"—not how, not by when, and not by whom.
- **Objectives**—These are the major activities or initiatives that the organization will conduct or complete that directly support a specific goal. These are more finite and more descriptive than goals but not as specific (yet) as an *action item* or *tactic* coming next. A good example of an objective that would support the goal of producing drinking water of the highest quality could be something like "participate in the Partnership for Safe Drinking Water Treatment Plant Optimization Program."
- **Action items**—Part of the implementation of the plan, action items are activities where one person a group actually does some defined work—with a specific date, often a specific budget, and with specific responsibility and accountability. For example, an action item to support participation in the Partnership for Safe Water might read like this: "By June 30, 2014, complete PSW Phase I application for the North River Treatment Plant, and begin gathering baseline data." To be complete, the action item would also name an individual or team that would be responsible for completing the task.
- **Performance measures or metrics**—Performance measures can be any of a number of metrics or indicators that help report and/or display that progress of some kind is being made toward the eventual achievement of one of the goals, objectives, or even action items identified in the plan. Taken most literally, such metrics are tracking the outcome or output from a process or behavior that was specifically identified as needing improvement or enhancement—or maybe just doing

something new and totally different from what you did before. The measures can come in many forms, but most importantly, they need to fit your particular situation and they need to matter. More on this later when tracking and reporting progress as a key aspect of plan implementation are discussed.

Types of Plans

In addition to some of the key vocabulary, it is also important to distinguish among different types of plans. Utilities are familiar with plans of all kinds, but it is worthwhile to define five particular plans: strategic plan, master plan, capital plan, operating plan, and business plan. One version of these is offered by a group that specializes in working with nonprofit organizations, foundations, and government organizations.

- **Strategic plan (or core plan)**—A tool that provides guidance in fulfilling an organization's mission with maximum efficiency and impact. The plan articulates specific goals and describes action steps and resources needed to accomplish them. Such plans are reviewed as frequently as annually and more thoroughly reviewed and revamped every three to five years.
- **Master plan**—In a utility a master plan usually documents medium-term and long-term plans for major projects, providing a comprehensive description of significant capital investments, the concepts that justify their inclusion, and their relation to urban design, infrastructure, service delivery, and regulatory requirements. It consists of images, texts, diagrams, statistics, reports, maps, and aerial photos that describe how specific projects will be developed. Ideally, it provides a structured approach and creates a clear framework for planning and decision making.
- **Capital plan**—Frequently focuses on budgeting resources for the future of a utility's large expenditures and long-term investment needs. Capital planning includes new and replacement machinery, structures, treatment plants, transmission networks, and water resource and reservoir protection needs as well as new plants, dams, and other major investments.
- **Operating plan**—A coordinated set of tasks for carrying out the goals in a strategic plan, the operating plan goes into greater detail than the strategic plan, identifying both time frames and roles/duties for individuals. Usually spans only one fiscal year.
- **Business plan**—This plan is focused on the actions and investment necessary to generate income from a specific program or service. It includes information about an organization's products, competitive environment, and revenue assumptions.

For our purposes, we will focus here on the strategic plan and the operating or implementation plan.

At an introductory level, it is perhaps best to look at the strategic planning process as a guide for evolving policies and programs consistent with the organization's core philosophy and a means to provide clear definition of those policies as staff-level actions that align with strategic objectives. In other words, we use this plan not only to align our values and mission and vision with strategic goals but also to organize our everyday efforts.

The implementation process (described in more detail in the following sections) is used first to frame and second to monitor and report progress on the steady, strategic, and continual achievement of stated goals and objectives—things that are important to living your mission and becoming what you envision. Through the development and internal implementation of the plan, each employee will know precisely what he or she must do (both as an individual and as an organizational team member) to support the mission and goals of the utility.

The strategic plan and the implementation plan both need to be clear and simply stated in terms that all staff recognize, relate to, and resonate with. These plans must also be readily explained and communicated to governance, community leaders, stakeholders, and rate-paying customers that you serve daily. As such, the plan and the overall planning process may even become tools for political leverage, community participation, and buy-in.

On occasion, the strategic plan will also be a mechanism to provide explanation or even justification for actions and decisions that need to be made as utility managers. One common experience/example is the situation when a utility manager finds himself/herself in a budget discussion with a board or council member, responding to a question about a line item that caught a member's eye (or maybe "ire"). How might that conversation have been improved if as a manager you could have provided this response: "Well, the reason we are requesting a new pump station is because it is one of the key action items in support of our water quality goal in the strategic plan that was approved and adopted last year."

Fundamentally, the strategic plan is a tool to guide the work of the utility. Accordingly, the plan development process will ultimately produce a set of actions that will direct management and help allocate resources in a manner that leads the utility toward the vision of the future it has defined, without sacrificing its ability to address its ongoing day-to-day business.

One perspective that relates well to utilities comes from the Department of Facilities Planning and Management (FP&M) at Iowa State University. In FP&M, they believe that their strategic planning process (and the plan it generates time after time) does the following:

- Forces consideration of the future and provides an opportunity to influence the future or assume a proactive posture.
- Provides better awareness of needs and of the organization's related issues and situation.
- Helps define the overall mission of the organization and focuses on objectives.
- Provides a sense of direction, continuity, and effective staffing and leadership.
- Plugs everyone into the system and provides standards of accountability for people, programs, and allocated resources.
- Is key to helping an organization collectively and cooperatively gain control of the future and the destiny of the organization.

STRATEGY AND STRATEGIC PLANNING: WHAT UTILITIES ARE DOING

Now that we have established some basic concepts and terminology, we can describe how the core strategic plan and the all-important implementation plan and strategy are created and launched. Remember, in a nutshell, the strategic planning process answers the following three questions:

1. Where are we today?
2. Where do we want to end up in the future?
3. How are we going to get there?

One common way to organize the plan development and implementation process is in three distinct phases:

1. Preplan phase
2. Core plan phase
3. Implementation phase

Phase 1: Preplan Organizational Assessment

This phase of the strategic plan development process varies widely from organization to organization. Many factors contribute to determining how complex or extensive the preplan assessment should be, but in every case close consideration should be given to the following:

- How much information you already have to assess your current situation
- How extensively you want to involve your staff and other stakeholders, including governance
- How much time and how many resources you have to contribute to this phase of the overall effort

Preplanning helps to establish a foundation for planning that recognizes external and internal drivers impacting the future of the organization/water utility. Regardless of what tool or technique you utilize, at the end of the process you will want to be able to answer the following questions:

- What are the strengths of our organization in its current state?
- What areas do we know or think we want to improve?
- What internal drivers or situations are calling us to make improvements?
- What outside pressures/external drivers are we facing that require us to change or improve?

Assessment Tools

With these questions in mind, a number of options are available to your organization to gather and evaluate the information that will enable you to build a solid strategic plan. One commonly used assessment tool is something called a *SWOT analysis* (Figure 3-2). Developed by Stanford University Business School as a tool to assess the feasibility of bringing a new product to market, the analysis is typically used in a facilitated workshop setting that encourages participants to brainstorm a list of strengths, weaknesses, opportunities, and threats (SWOT) around a particular topic or theme. With water utility organizations, the tool is more useful if you have already identified a list of key issues or drivers or even future possibilities. That way, each issue or possibility can be assessed, and a resulting set of opportunities can be aggregated and prioritized for future consideration as part of the strategic plan.

Another version of the SWOT that helps utilities look at different aspects of a particular situation or issue is known as the *PEST analysis*. With this tool, each issue or situation is assessed from four perspectives—political, economic, social, and technological. With either SWOT or PEST, the goal of the exercise is to assess outside or external realities facing the organization, and to give consideration to those realities when identifying a strategic direction as an organization.

It is also important to assess the organization from the inside, seeking to identify both the positive aspects/strengths and the areas that might need improvement. This assessment can be done in a number of ways, including the use of individual and small group interviews and/or the use of a survey tool.

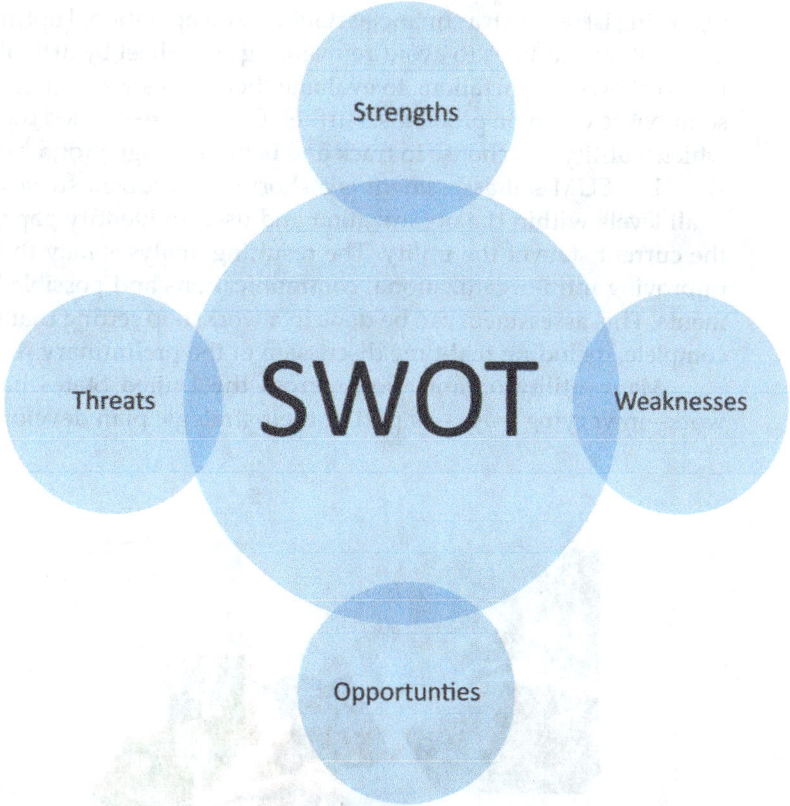

Figure 3-2 Components of SWOT analysis

A variety of resources exist to support this kind of evaluation, including a number of documents and tools listed at the end of this chapter. One obvious yet frequently overlooked organizational improvement resource is the well-facilitated event or meeting. By understanding the group's desired outcome and the background and context of the meeting or event, a facilitator will go on to (1) design and plan the group process and select the tools that best help the group progress toward that outcome and (2) guide and control the group process to ensure that

- There is effective participation.
- Participants achieve a mutual understanding.
- Participants' contributions are considered and included in the ideas, solutions, or decisions that emerge.
- Participants take shared responsibility for the outcome.
- Outcomes, actions, and questions that come from the process are properly recorded and acted on, and appropriately dealt with afterward.

One of the popular organizational self-assessment tools used recently is EUM, which stands for *effective utility management* and was developed as part of a multi-association effort (see Figure 3-3).

The EUM program is a set of tools based around 10 attributes of an effectively managed utility and five keys to management success (see Figure 3-4). EUM attributes and terminology provide a language through which utilities across the nation can communicate regarding issues such as financial stability and operational optimization. This common language allows utilities to avoid reinventing the wheel by articulating industry challenges. It also allows organizations to evaluate themselves versus peers, as well as share ideas or solutions to common problems. Further, EUM has expanded the performance metrics from which a utility can choose to track and better manage progress toward benchmarks.

The EUM self-assessment is a short standardized survey that can be administered at all levels within the organization and used to identify gaps in perspectives regarding the current state of the utility. The resulting analyses may then serve as focal points for improving intra-organizational communications and possible business process improvements. This assessment can be done in a workshop setting that takes two or three hours to complete, including real-time discussion of the preliminary results.

Many utility organizations across the United States have used the EUM framework—in varying ways—as part of their strategic plan development and implementation

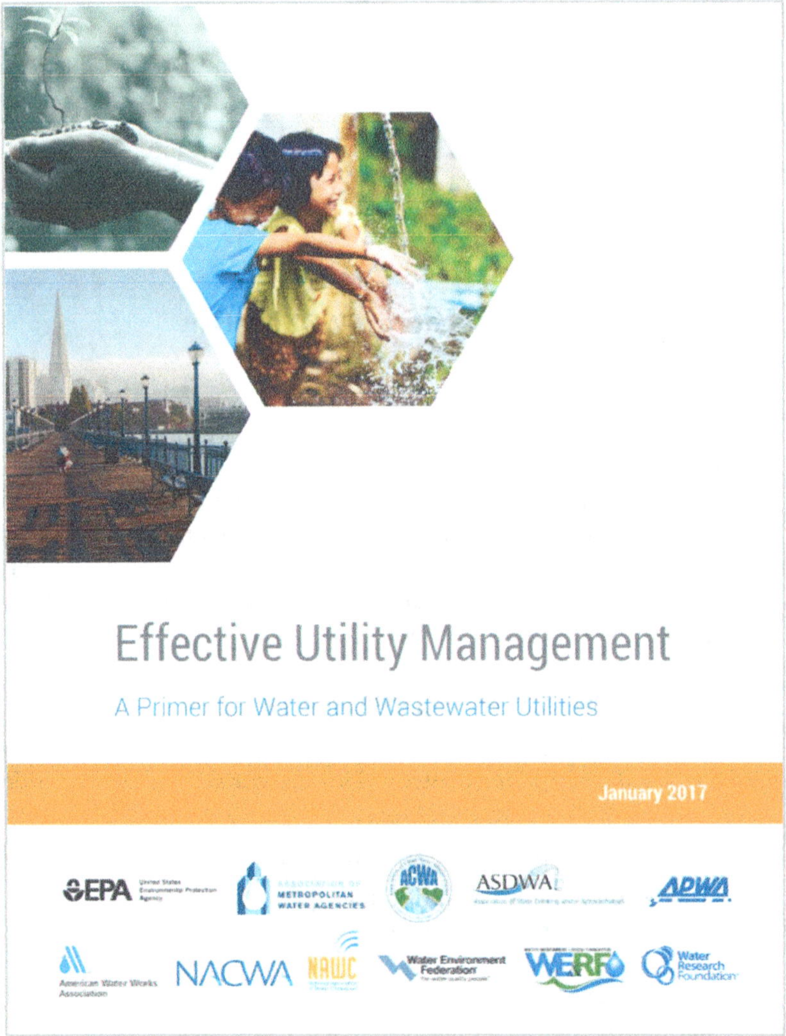

Figure 3-3 Multi-association development of *Effective Utility Management Primer*

Source: EUM Primer USEPA 2017

Figure 3-4 Ten attributes of effectively managed utilities and five keys to management success

Source: EUM Primer USEPA 2017

processes. Some of these include DC Water, the Birmingham Water Works Board, Central Arkansas Water, and the City of Newark (Ohio). Links to these organizations are provided later in this chapter.

Phase 2: Strategic Plan Development

Based on the findings and discussions from the preplanning assessment phase, the organization is now ready to proceed with the development of the core plan (or strategic plan). Once completed, the core plan will contain five components:

- A vision statement
- A mission statement
- A values statement or list of values
- A set of strategic goals
- A set of objectives for each of the strategic goals

Once again, the process that is followed to develop each of these core components varies from utility to utility. Some utilities have professional staff that possesses the experience and skills to assemble a committee or a working group and then work with that group to build the plan. Other organizations enlist the help of outside facilitators, but in all cases the most important aspect of the process is the development of the committee or working team that takes on the responsibility to represent the rest of the utility staff in the process. Selection of the team members is an important task for utility leadership and great care should be taken to ensure that the staff is well represented, both in breadth and depth.

Phase 3: Implementation of the Plan

With the core plan in hand, the work of the strategic planning committee or working team has just begun. The real work starts with the implementation of the plan, and the implementation phase begins with the development of the last two building blocks of the strategic planning process—the list of action items and a set of performance measures. In common terms, this is where the rubber hits the road and the actual changes that lead to progress and achievement take place. Overall, the implementation phase has five key components:

1. Developing and prioritizing action items
2. Identifying metrics and indicators to track and report progress
3. Performing the work prescribed in the plan
4. Committing to communicating progress
5. Revisiting and revising the plan on regular basis

Action items. Action items are specific and measurable activities that achieve an objective and that identify both a responsible party and a timetable for implementation. Importantly, these should be actions or activities that are consciously identified (and assigned) to either enhance existing practices and processes or, in some cases, to add to what is already in place. As such, this may mean actually replacing or eliminating established activities or practices. Accordingly, it is important to have adequate discussion and concurrence among utility staff before enacting these parts of the plan.

Performance measures: Metrics and indicators to track and report progress. A key ingredient to successful implementation of the strategic plan is the identification of specific metrics or indicators that enable the utility to track and report its activity and/or performance related to making progress on the achievement of its stated goals or objectives. There are a number of ways to do this—including formal processes like the balanced scorecard (Kaplan and Norton 2001; Figure 3-5) or customized performance management systems that fit the particular nature and needs of the utility. Please see the sources listed later in this chapter for examples and links to some of the available and often used options.

Regardless of which way you go, it is fundamentally important that whatever measures, metrics, or indicators you identify have the following characteristics, namely, that they are (1) specific and clearly understood; (2) measurable and quantifiable; (3) achievable; (4) relevant (especially with regard to linking to the plan goals and/or objectives); and (5) time based (meaning they have a targeted due date).

Doing the work. Once the core plan has been built and the corresponding action items and performance measures or indicators of success have been established, work to implement the plan begins. Once again, there are many ways to implement the process. Keys to getting the work done include

- Assessing and committing to the right level of effort and adequate resources (especially time) to implement the plan, including prioritization of action items and the implementation of some kind of performance management system

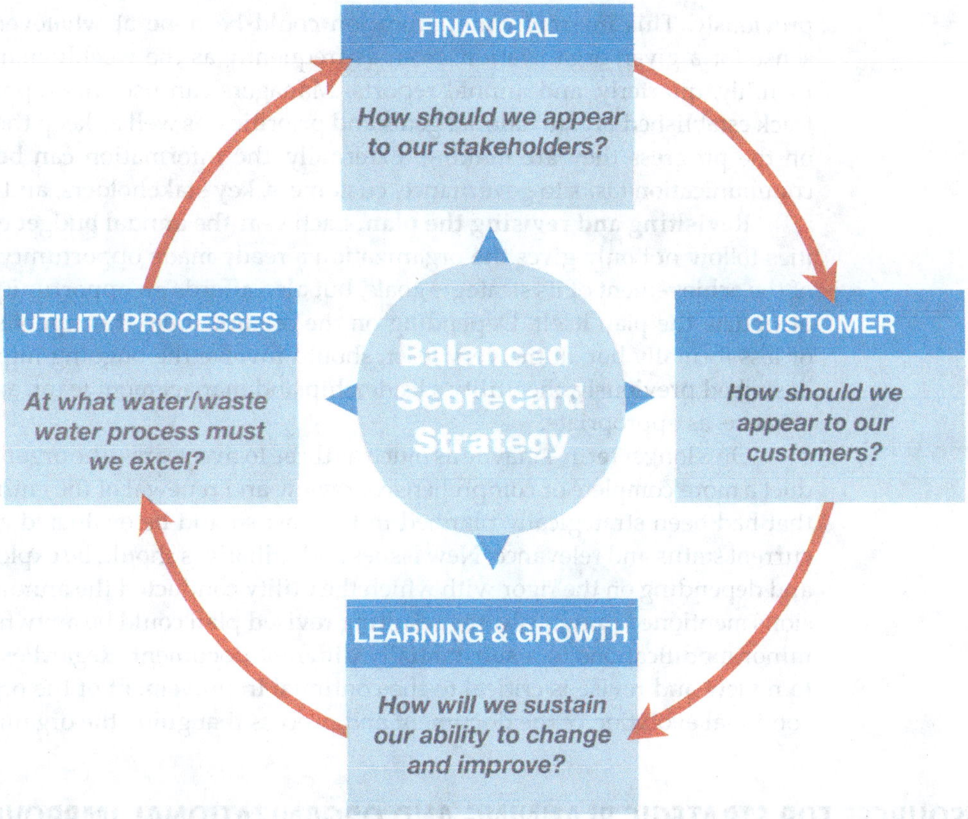

Figure 3-5 Adapting a balanced scorecard approach

Source: Kaplan and Norton 2001.

- Building and mobilizing goal-based implementation teams that not only do the work identified in the action items (which are linked to each of the strategic goals in the plan) but also take responsibility for tracking and reporting progress on the achievement of the strategic business goals
- Connecting the work of the teams to the overall work of the organization, including informing the utility management team on a regular basis so it can monitor progress and report that both across the organization and outside the organization

Communicating progress regularly. Every successfully implemented strategic plan has a corresponding communication strategy/plan. Part of the implementation process is a communication planning session/workshop that includes

- Discussing the purpose and function of the communication plan
- Identifying communication goals and objectives, including details related to content, format, and frequency
- Identifying audiences (internal and external)
- Identifying the range of potential communication methods (meetings, email, newsletter, web sites, etc.)

Typically, the communication strategy/plan for the strategic plan will have two areas of focus: internal communication and external communication. Internally, the process includes the regular reporting of the established measures and metrics mentioned previously. This internal communication could be done at whatever frequency makes sense for a given organization, from as frequently as the weekly managers' meeting, to monthly, quarterly, and annual reports. Managers can use this reporting to help them track established organizational goals and priorities, as well as keep the entire staff posted on the progress they are making. Externally, the information can be linked to specific communication tasks to governance, customers, key stakeholders, and the general public.

Revisiting and revising the plan. Each year the annual budget cycle that most utilities follow not only gives the organization a ready-made opportunity to assess progress of the achievement of its strategic goals, but also affords an opportunity to modify, adjust, or update the plan itself. Depending on the organization, this update can be done more or less formally but, at the very least, should involve the ongoing implementation teams described previously, the utility's leadership and management team, and the utility's governance, as appropriate.

On a longer term—maybe as much as three to five years—the organization should conduct a more complete or comprehensive review and renewal of the entire plan. Everything that had been strategically planned in the past should be evaluated and assessed for its current status and relevance. New issues and initiatives should be explored and discussed, and depending on the rigor with which the utility conducted the annual reviews and revisions mentioned previously, the resulting revised plan could be anywhere from a series of minor modifications to a substantially different document. Regardless, the commitment to review and revise is critical to the continual improvement of the organization and the continual evolution of the document and process that guide the organization.

RESOURCES FOR STRATEGIC PLANNING AND ORGANIZATIONAL IMPROVEMENT

As mentioned previously, there is no one way to build and implement a strategic plan. Yet, regardless of how you get it done, it is a more and more important component of a well-managed water sector utility.

It is hoped that this chapter serves the dual purpose of highlighting the importance of having a strategic plan and providing some basics of a plan that can work for utility organizations like yours. To help you get started or perhaps to breathe new life into an outdated and never-really-implemented plan on your shelf, please consider the following resources (organized by topic).

Strategy and Strategic Planning Resources

- *The Strategy-Focused Organization,* Robert S. Kaplan and David P. Norton, Harvard Business Review Press, Brighton, Mass., 2001. From the creators of the performance management tool called the *balanced scorecard,* this book introduces a new approach that makes strategy a continuous process that involves both managers and staff at all levels of the organization.
- **On Strategy**, Harvard Business Review Press, Brighton, Mass., 2011. One in the series of HBR's "Ten Must Reads," this is a collection of definitive articles on strategy and strategic planning.
- **Association for Strategic Planning (ASP)**—A North America–based organization that holds national conferences, supports a number of state/province-based chapters, and importantly develops standards and best practices for strategic planning practitioners, ASP also has certification programs for both individuals and organizations. Find more information on ASP's web site at www.strategyassociation.org.

For example, here is the Association for Strategic Planning's list of "Top Ten Best Practices for Qualifying Strategic Planning and Management Frameworks":

1. Use a systems approach
2. Incorporate change management and leadership development
3. Provide actionable performance information
4. Incorporate assessment-based inputs
5. Include strategic initiatives
6. Offer a supporting toolkit
7. Align strategy and culture
8. Integrate existing organization systems and align the organization around strategy
9. Be simple to administer, clear to understand and direct, and deliver practical benefits over the long term
10. Incorporate learning and feedback to promote continuous long-term improvement

- **Department of Facilities Planning and Management (FP&M) at Iowa State University.**

  http://www.fpm.iastate.edu/about.asp

- **The TCC Group—with offices in New York, Philadelphia, and Chicago.**

  Web site: www.tccgrp.com Email: info@tccgrp.com

  For over two decades, TCC has provided strategic planning, program development, evaluation, and management consulting services to non-profit organizations, foundations, corporate community involvement programs, and government agencies.

  For consideration, here is a list of TCC's "Ten Keys to Successful Strategic Planning for Non-Profit and Foundation Leaders":

1. Clear and comprehensive grasp of external opportunities and challenges
2. A realistic and comprehensive assessment of the organization's strengths and limitations
3. An inclusive approach
4. An empowered planning committee
5. Involvement of senior leadership
6. Sharing of responsibility by board and staff members
7. Learning from best practices
8. Clear priorities and an implementation plan
9. Patience
10. A commitment to change

Benchmarking and Performance Improvement Resources

- **Benchmarking Performance Indicators (from AWWA Utility Benchmarking Survey).** Since 2004, AWWA has conducted an annual survey to collect utility benchmarking data from water and wastewater utilities. AWWA's Utility Benchmarking Program tracks recognized utility performance indicators developed and applied by water industry professionals to provide objective performance measures for decision makers responsible for a utility's continual improvement. The annual survey produces a final industry report with performance indicator results for water, wastewater, and combined utilities for the fiscal year. Visit www.awwa.org/benchmarking to participate in the current survey or to find more information on the latest industry report.

To make valid comparisons, performance indicators must be well defined and consistently used in context. If definitions are inconsistent or incomplete, the resulting performance data will not be comparable. Even when comparable data are collected, external comparisons should be taken into consideration because numerous system-specific factors can influence the system performance.

- **Total Water Management—Industry Survey Builds on 10 Attributes of Effective Utility Management,** Hasit, Yakir; Yeager, Fair; and Matichich, Mike, AWWA's *Opflow,* September 2012, Vol. 38, No. 9.
- The Water Research Foundation report *Performance Benchmarking for Effectively Managed Water Utilities* addresses qualitative metrics by using process or performance benchmarking, with each process focusing on a utility's capability regarding the practice areas identified in each of the 10 attributes of effective utility management (EUM) identified in *Effective Utility Management: A Primer for Water and Wastewater Utilities,* compiled by industry organizations and the US Environmental Protection Agency (USEPA). Key metrics and practice areas for achieving the 10 attributes were identified through literature review, water/wastewater utility participation, and development of a performance benchmarking framework and tool for each attribute. In addition, the framework, methodology, assessment, and tool for EUM benchmarking were pilot tested with water and wastewater utilities. Participating utilities will be able to use benchmarking to track their progress and benchmark themselves against leading practices to develop improvement plans.
- **Canadian National Benchmarking Initiative.** The initiative was started in 1998 and has since grown to 55 member municipalities participating in stormwater, water, and wastewater benchmarking. This project was developed in response to a need for Canadian municipal water and wastewater utilities to measure, track, and report on their utility performance. While fundamentally a high-level metric benchmarking process, it has developed into a network and information base for Canada's most progressive municipal utilities. Their web site presents information about the project and it is the focal point of project-related communication for member municipalities ("Member's Area" section). In the context of this initiative, benchmarking is the ongoing process of comparing products, services, and practices with those of similar organizations. The ultimate goal is always to improve quality and performance. For more information and to assist researchers and others who are interested, the partnership has made available some of the blinded data in a min/max/avg graph format for stormwater, wastewater, and water, all accessible at the Public Report link (http://www.nationalbenchmarking.ca/).
- ***Benchmarking Water Services: Guiding Water Utilities to Excellence***, Enrique Cabrera Jr., Peter Dane, Scott Haskins, and Heimo Theuretzbacher-Fritz, The Netherlands, IWA Publishing, 2011.

General Improvement and Quality Resources

- **AWWA Utility Management Standards**
  http://www.awwa.org/store/standards/utility-management.aspx. The following is a list of some AWWA standards useful in the management context.
  - AWWA G100, Water Treatment Plant Operation and Management
  - AWWA G200, Distribution Systems Operation and Management
  - AWWA G300, Source Water Protection
  - AWWA G400, Utility Management System
  - AWWA G420, Communication and Customer Relations
  - AWWA G440, Emergency Preparedness Practices
  - AWWA G480, Water Conservation Program Operation and Management
  - AWWA G481, Reclaimed Water Program Operation and Management
  - AWWA G510, Standard for Wastewater Treatment Plant Operations and Management
- **Effective Utility Management (EUM) Initiative and Related Resources.** Since 2008, USEPA has been working collaboratively with six national organizations that support drinking water and wastewater utilities to promote effective utility management (EUM) based on a series of attributes of effectively managed water sector utilities (attributes). Much of this work has been targeted at medium-size and larger utilities that have seen significant benefits from using EUM to improve their operations. The primer was updated in 2017 and is more relevant than ever before to help meet the challenges facing utilities. EUM is a starting point for any utility's path to effective and sustainable operations. Additional information about this initiative is located at www.watereum.org.
- Some examples of utilities using EUM as part of their strategic plan development and implementation process:
  - **DC Water, Washington, D.C.**
    - http://www.dcwater.com/news/strategic.cfm
  - **Birmingham Water Works Board, Birmingham, Ala.**
    - https://www.bwwb.org/strategicplan
  - **Central Arkansas Water, Little Rock, Ark.**
    - http://www.carkw.com/2012/10/central-arkansas-water-recognized-for-management-excellence/
  - **City of Newark, Ohio, Division of Water and Sewer**
    - http://www.newarkohio.net/city-services/water–wastewater-main
    - **Using EUM to Broaden the Scope and Depth of Strategic Planning,** AWWA/WEF Utility Management Conference, Miami, Fla., 2012
- **American Productivity and Quality Center (APQC)**—APQC is a member-based nonprofit and one of the world's leading proponents of business benchmarking, best practices, and knowledge management research. APQC's mission is to help organizations around the world improve productivity and quality by *discovering* effective methods of improvement, broadly *disseminating* findings, and *connecting* individuals with one another and with the knowledge they need to improve.
  https://www.apqc.org/about

This page intentionally blank.

Chapter 4

Workforce Issues: How to Get, Grow, Keep, and Replace the Best Staff Possible

This chapter addresses topics across the spectrum of workforce management, including recruiting, capability development, employee retention, and sustaining knowledge within utilities. The purpose is to provide water and wastewater utility managers the basis for a proactive management approach to the people side of the business.

In 2010, the Water Research Foundation (WaterRF) and AWWA commissioned the *Water Sector Workforce Sustainability Initiative* ("the report," Brueck et al. 2010). The report found that the exodus of baby boomers from the workforce began in 2005 and will probably continue through at least 2025. The authors estimated that before 2020, the industry is likely to lose between 30 and 50 percent of the current workforce. In addition, the report concluded that the positions most likely be affected by the impending retirements are in engineering and water operations. Other organizational areas, such as those requiring specific technical operational knowledge, like SCADA, also are impacted.

For many years, water utilities have relied on a pool of knowledgeable and technically competent workers from which to draw new employees. Increasingly, utilities have, however, experienced shortages of qualified candidates for critical positions becoming vacant as a result of the impending retirement of members of the baby boomer generation. The water industry is competing more intensely with other industries for a shrinking pool of qualified employees.

Utility executives seeking to replace employees are also challenged to find additional staff to handle increased regulatory requirements, the technological complexity of water systems, growth, infrastructure replacement, customer demands, and increased security. Compounding the workforce issue is the fact that many of these same executive managers and employees in leadership positions are part of the generation retiring. As a result, utilities are also faced with the challenge of developing their next generation of supervisors, managers, and leaders.

The data noted in the 2010 report are not the only source of critical data on workforce issues. Since 2007, AWWA's annual State of the Industry surveys (Mann and Runger 2007, 2008, 2009, 2010) have cited workforce matters as one of the top five priorities. In the article "Top 10 Jobs That Americans Can't Live Without," *Reader's Digest* (Stockdale 2012) ranked water and wastewater treatment plant operators and system operators as the number two essential and most secure professions, topped only by registered nurses. Clean and safe potable water is necessary for life.

Utility managers can refer to strategies outlined in the following sections to fill critical vacancies so that they may continue to serve their customers as public health providers.

UNDERSTANDING WORKFORCE ISSUES

Evaluating Your Workforce Programs

Workforce issues can be defined as anything that impacts a utility's ability to recruit, manage, develop, and retain the employees it needs to be successful. There are several workforce evaluation frameworks and two are included in Figures 4-1 and 4-2. Workforce issues can be viewed from the perspective of the utility (Figure 4-1) or the perspective of the employee through the stages of the employee life cycle (Figure 4-2).

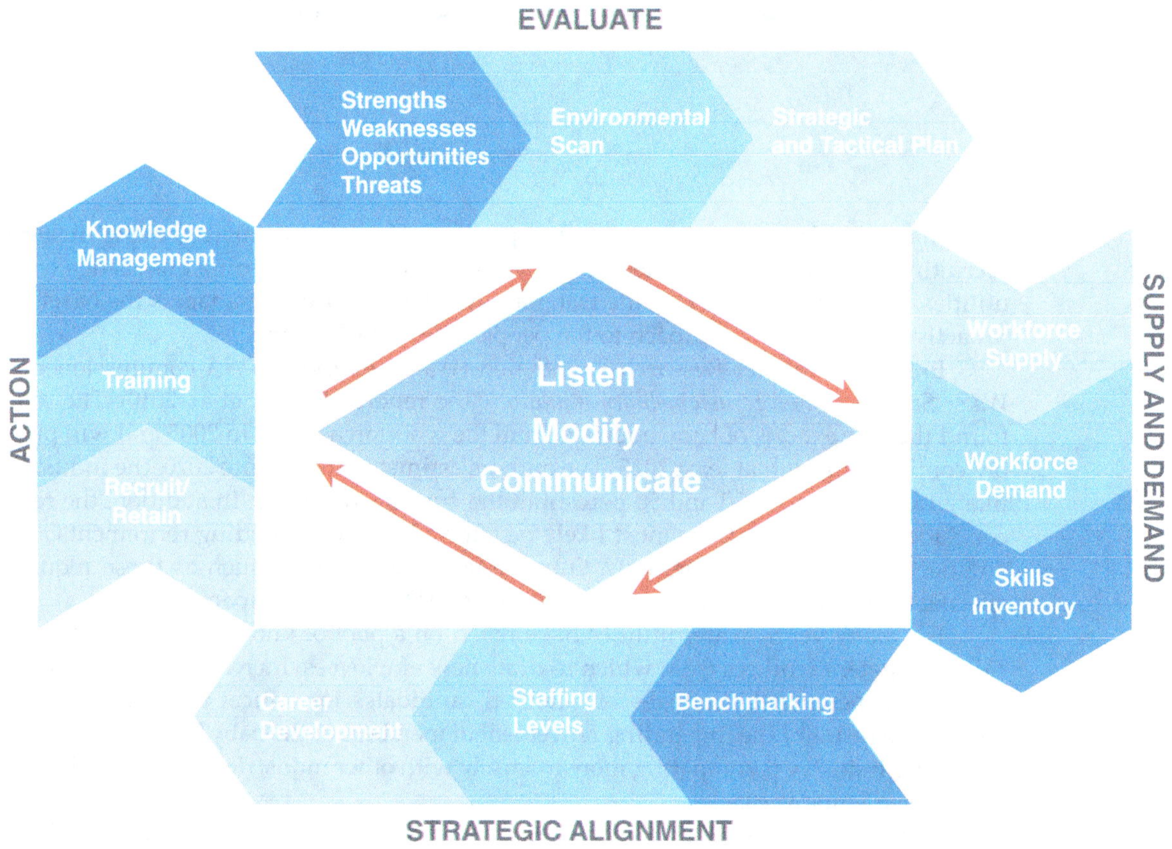

Figure 4-1 Workforce utility planning model

Source: Olstein et al. 2005. Reprinted with permission of AwwaRF and Water Environment Research Foundation.

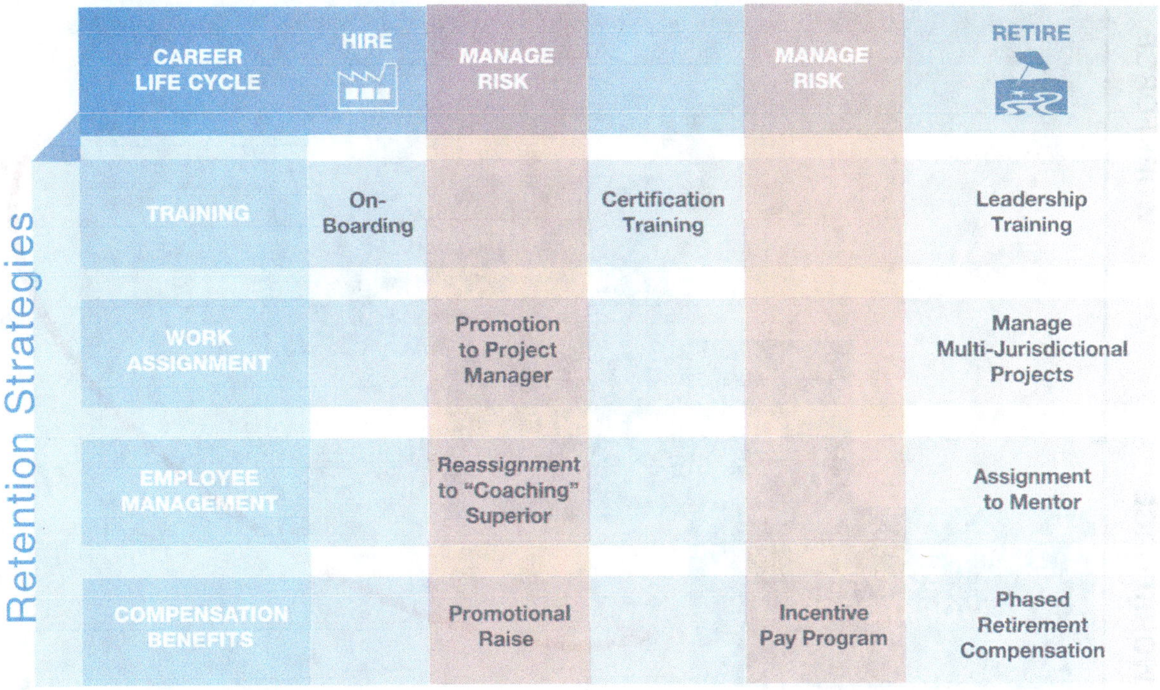

Figure 4-2 Employee life cycle

Source: Brueck et al. 2010. Reprinted with permission of WaterRF.

Each figure depicts the most critical stages in planning and developing a utility's workforce. Figure 4-1 is broken into four major phases of the workforce planning cycle. Note that the utility has control of three out of four of these phases: scan, optimize, and action. Only the supply and demand phase is largely controlled by external forces such as economic and geographic conditions. Utilities can implement several programs and techniques to ensure effective growth and retention of its current workforce. This is an important step toward overall sustainability of a skilled workforce.

Figure 4-2 highlights critical milestones in an employee's career development. By incorporating these growth milestones into an overall workforce planning model (Figure 4-1), a utility can improve the likelihood of employee retention, thereby reducing the impact of external forces. Workforce supply and demand can never be controlled by a utility; however, through effective planning, utility managers can mitigate the risks involved.

Workforce issues differ in their effect and importance by location, community size, and socioeconomic climate. They also differ based upon various functions and job types within each utility (Manning 2008). Each utility should conduct its own analysis, following the framework, illustrated in Figures 4-1 and 4-3 to determine how prepared it is to manage its own workforce issues.

Water Industry Trends

In the report initiated by WaterRF in 2010 (Brueck et al. 2010), responses to workforce issues were categorized into national, regional, and local-level responses within the water industry. Responses were then further categorized into opportunities for collaborative approaches with other industry sectors and other types of partners, for example other government organizations or educational institutions (Figure 4-4).

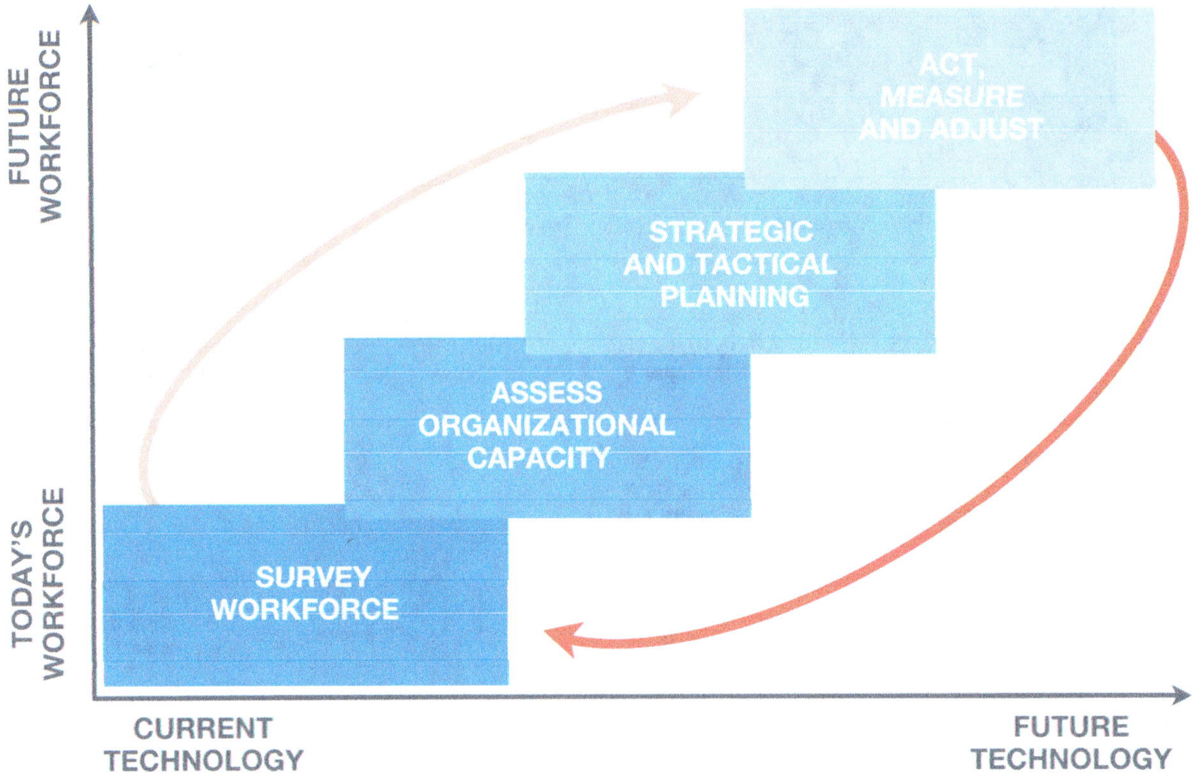

Figure 4-3 A four-step process used to address utility workforce needs

Source: Manning 2008. Reprinted with permission from AWWA Research Foundation (now WaterRF) and US Environmental Protection Agency.

Individual utility responses have included partnerships with schools and colleges and outreach to veterans and other displaced workers. These partnerships have involved identifying candidate pools, starting internship programs, identifying recruitment and retention strategies, succession planning, mentoring, utility-specific training programs, development of a water-industry job competency model, and many other initiatives.

Utilities are encouraged to contact peer organizations, local AWWA sections, and other industry associations, and to consult the references listed in this document for case study examples relating to their own workforce planning.

Examples of regional collaboration include

- Baywork (http://baywork.org), a consortium of San Francisco Bay–area water and wastewater utilities
- Get Into Water (http://getintowater.org), an alliance of Colorado-based utilities, workforce centers, educational institutions, and associations
- Work in Water consortium (http://workinwater.org) based in Arizona
- H2opportunity (http://h2opportunity.net) in Georgia

The number of similar programs continues to increase. All of these sites represent an excellent starting point for utilities and their managers as they begin to address their own specific workforce issues. Utilities are encouraged to contact their local AWWA or WEF sections for additional examples.

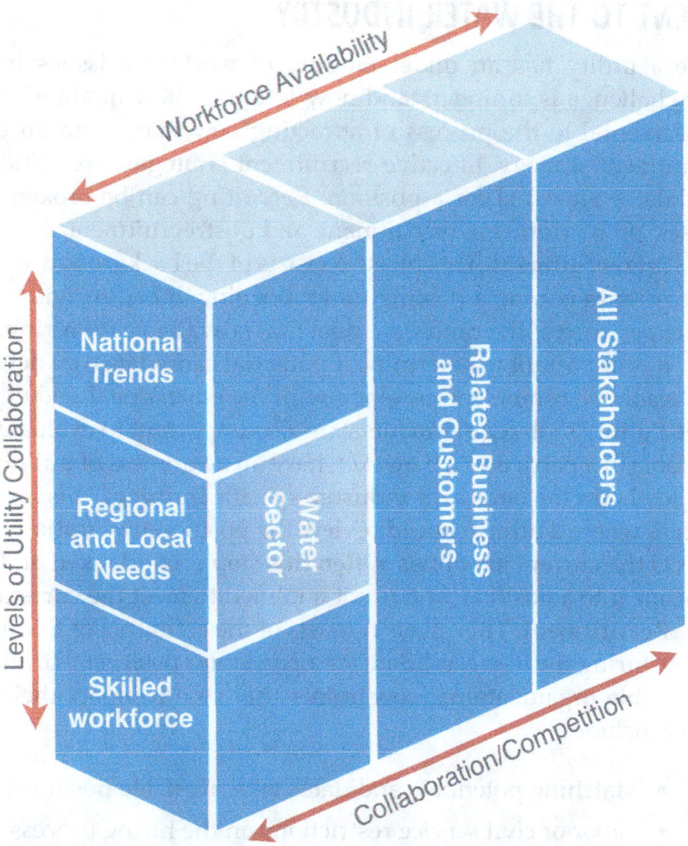

Figure 4-4 Framework for collaboration to address workforce issues

National Trends

As noted earlier, it has been projected that 37 percent of water utility workers and 31 percent of wastewater utility workers will retire within the next 10 years. Utilities will feel the impact of these retirements most severely in job classifications requiring technical skills and knowledge such as engineering, plant operations, water quality, and SCADA.

The US Bureau of Labor Statistics has estimated that the water supply and sanitary services sector will experience a 45 percent increase in the need for additional employees in coming years because of new regulations, infrastructure growth, security challenges, and customer demands.

Regional and Local Needs

The shortage of skilled workers is more severe than noted in the Water Research Foundation report (WRF 2012) for plant operators in the Northeast and the West. Plant operators in the Northeast (average 49.6 years old) and in the West (average 46.9 years old) are significantly older than plant operators in the Midwest (average 41.2 years old) and the South (average 40 years old). Similarly, plant operators in metropolitan areas (average 42.4 years old) are older than plant operators in nonmetropolitan areas (average 36.9 years old). This reality makes succession planning—at least with regard to plant operators—a slightly more pressing issue in the Northeast and West than it is in the South and Midwest regions. It is also more of an issue for utilities in metropolitan areas than in nonmetropolitan areas.

ATTRACTING TALENT TO THE WATER INDUSTRY

Once a utility has an understanding of workforce issues in the water industry, the next challenge is to understand how to attract new qualified talent to the organization. Fundamental to the process of attracting new talent into an organization is forming a recruitment strategy. Effective recruitment strategies are critical to ensure that the best candidates are hired for a position. Recruiting can be broken down into three primary phases: prerecruitment, recruitment, and postrecruitment.

Prerecruitment. This phase consists of the background work that utilities must complete prior to issuing a posting for an opening. It begins with a review of the knowledge and experience of the candidate pool that is available for hiring.

It is important to determine if potential candidates are developing the basic personal and academic competencies, along with the educational and experience qualifications, to meet a utility's job requirements. It is also important to evaluate if suitable candidates are available in a particular job market, have an awareness of utilities as a career path, and are able to obtain the necessary industry-specific training. This is also the point in the hiring process where a utility should review the position description to ensure that it accurately reflects the current job duties. Often this step is overlooked and the result is the hiring of someone into a position who is not qualified to meet the current requirements of the job.

Recruitment. This phase consists of the actual act of finding, interviewing, and ultimately hiring the best candidate for a particular position. Hiring managers must be aware of possible organizational constraints that could impact their selection process. Likely issues include

- Matching potential candidates with available positions
- Labor or civil service restrictions on the hiring process
- Residency requirements within municipal jurisdictions
- Ability to compete with private-sector compensation and benefits packages

In every situation, the hiring manager must remain determined through this phase to ensure that the right candidate is hired for the position. Oftentimes, when a manager feels pressured to settle for a candidate who is not the best fit for the position, performance issues result and a much lengthier and time-consuming separation process may be the outcome.

Postrecruitment. Once an employee is hired, there are workforce issues related to orientation, training and development, advancement opportunities, and performance recognition that need to be addressed. Addressing these issues helps to ensure that individual employees have the skills and abilities they need to perform their work successfully. Knowledge management, including the capture, storage, and transfer of institutional knowledge between departing and new workers, is critical and should be incorporated into a utility's coaching program to ensure transfer of knowledge to new employees.

Recruiting Challenges

The three primary challenges to recruiting in the public water–wastewater industry are (1) lack of qualified workers, (2) lower salaries, and (3) prestige. Additional challenges include the training, certification, and licensing of new employees and the loss of industry knowledge as an aging workforce retires. Each of these barriers can be a major deterrent for potential candidates to move into the industry. When more than one barrier is present, many talented candidates will just choose a different profession. Further compounding the challenges is the fact that utilities are operating in a rapidly changing landscape with increased regulation, more demanding public expectations, and continually changing technology, processes, and equipment. These dynamic changes all impact the recruiting process.

Recruiting difficulties are more acute when dealing with operations positions. It is important for utilities to attract more highly educated water operators as system automation becomes the industry's future. Utilities must adapt to 21st-century technology, which requires more computer proficiency from new hires. Generally, the water plant operator career path has not been as desirable as other positions in the water industry, even as pay and benefit packages improve. Therefore, the water operator position will likely need to evolve into a professional position in order to attract new operators.

Additionally, economic conditions since 2008 have restricted the ability of many utilities and municipalities to hire staff. Instead of hiring and retaining adequate operator workforce, the trend has been to decrease the existing number of operator positions by attrition and reduction in forces. This downward trend is leading to a serious deficit in qualified staff available to operate and manage critical water systems, increasing the potential for employee burnout.

Salary Inequity Challenges

The issue of national salary inequity looms large when hiring future operations and maintenance personnel. An ongoing, erroneous industry perception is that operations and maintenance staff are not "professionals," which has created a significant barrier to attracting young people who are entering the workplace. The sophistication and complexity of the equipment, facilities, and systems require that the operations and maintenance staff be very technically competent; however, the salary structures typically do not support that proficiency. Utilities and municipalities are competing with private industries for qualified staff.

Distribution operator salaries for the highest level of large municipal treatment plant operators or lead operators with 15 to 30 years of experience have ranged between $50,000 and $62,000 per year. However, small system operator salaries are half to one-third of a large municipality due to the small number of ratepayers and in many cases the fact that there is only one operator.

For example, approximately 91 percent of the 1,650 operators in Montana hold three certifications and operate very small systems. The complexity of the job for a small system operator in a community with a population of 900 people may be significantly less, but the demands of the job are vastly increased by comparison to a large community with a population of 100,000 or more.

Many factors must be considered when evaluating the significance of pay inequity for a particular utility. These include job competencies and duties, certification levels, education, years of relevant experience, and job benefits. Additional factors include cost of living, geographic location, and whether the position is a private company or a public agency like a district or municipality. The inequality lies more in the demands of a particular job in the area where the salaries are compared. Utilities can utilize tools such as industry salary surveys to help determine what the correct salary for a particular position should be. Many of these surveys can be found in a utility's local region and can help establish a pay range that is more in line with other utilities or comparable industries in the area.

Availability of Technical Training

National trends indicate a decline in the number of adequately trained operators in both treatment facilities and distribution systems. The availability and affordability of technical training are also emerging as significant issues. There are several technical training programs throughout North America; however, the programs tend to be expensive and require a significant time investment of one to two years. This usually only results in a base level operations or maintenance license, and many utilities and municipalities do not recognize the value of the training. Additionally, due to the hiring requirements, it may be difficult to hire an operations or maintenance professional into a mid-level or higher position from outside the organization, which limits the opportunities for people who have invested significant time and money into getting the certificate or degree. This situation reinforces the notion that operations and maintenance positions have historically not been regarded as professional positions, which must be addressed to solve the larger hiring issue.

Fortunately, resources are available to develop and provide the appropriate training and support for someone who wants to get into the water industry. Professional organizations such as AWWA and WEF and their respective sections and member associations have several training programs and short courses available to members. Programs at various community colleges and trade schools both regionally and locally provide theoretical and hands-on training for many jobs in the utility profession.

The key to the success of these programs often is in how well they are publicized. Therefore, development of a clearinghouse type of web site that has links to the variety of programs available may be useful in getting the word out. Other programs to develop the opportunities at the high-school level are very useful to let young people know about the industry and the opportunities available.

Technical training for operators is available through a variety of resources, including AWWA textbooks, workbooks, and reference guides; state operator certification program web sites (https://www.epa.gov/dwreginfo/information-about-public-water-systems) via the US Environmental Protection Agency; California State University, Sacramento, correspondence courses in drinking water, wastewater and utility management; state environmental training centers; water conferences; private industry; technical assistance providers; and online distance learning programs.

A state's operator certification program must approve technical training programs needed for operator certification exams or continuing education requirements before credit can be awarded. There are numerous training programs that may be preapproved by your state. Certification criteria vary for each state. You should consult your local AWWA section for more information on training available in your area.

Impacts of State Licensure and Reciprocity

Reciprocity among states for an operator's license/certification is a significant issue. Very few states easily allow reciprocity and those that do often have challenging requirements. This lack has become a growing problem in an increasingly mobile society. If operations or maintenance professionals have worked their way up the ladder at their organization and are working in a mid-level or higher position, they are severely restricted in their ability to move to another state. As a result, experienced and qualified operations or maintenance professionals often have to start over again at the lowest position in a new utility if they choose to relocate.

State licensure requirements have vast public health implications that require water systems to retain certified operators. Utilities can find themselves in legal violation for failure to hire and retain certified operators. However, because licensure requirements vary significantly from state to state, reciprocity can be hard to achieve. Several states use standardized exams for some classification levels that may or may not be reciprocal with other states that have similar classification levels. A smaller number of state programs create their own exams or hybrid exams, which use standardized questions but may have state-specific questions added that make reciprocity harder to achieve.

The reciprocity becomes even more challenging for small communities that must find an operator or candidate with a background in areas such as chemistry, biology, math, engineering concepts, hydraulics, and public health who also is willing to be on call 24/7,

365 days a year. New technologies and greater operator performance expectations are making small-system operator compliance harder to achieve in the face of increased regulatory parameters.

Overall, operators find it difficult to achieve reciprocity in a state different than the state in which they are certified. Unlike professional engineering certification, which utilizes a national umbrella organization (National Council of Examiners for Engineering and Surveying [NCEES]) to standardize licensure, water industry operator certifications vary greatly from state to state. Classification categories, need-to-know criteria, experience, and education requirements also vary between state certifications. Fortunately, the water industry has begun to move, albeit slowly, toward standardization of all exams or professional status with a combination of exam, experience, and education submitted to a certification board.

Regional and Local Needs

Skilled worker shortages may be more acute in some regions and localities. The specific organizational cultures of utilities and municipalities can exacerbate the issue based on hiring and promotional practices. Capital funding is being released for replacement of facilities and systems that have reached the end of their useful lives, but capital funds do not pay for staffing. In addition, many of the local or regional trades and technical training centers that were available for preparing operations staff to work in this industry have been closed or dropped. These closures have led to a shortage of trained operations staff ready to step into the staff slots needed to operate and maintain the new, more sophisticated facilities and systems.

EMPLOYEE DEVELOPMENT AND SUCCESSION PLANNING

One of the best ways to mitigate the effects of a lack of adequate workforce supply is to grow and develop the employees within your utility. The cost of replacing an employee when he or she leaves is significant and can be avoided by investing in a utility's existing staff. Additionally, institutional knowledge is retained and less effort is spent on bringing a new employee up to speed on the requirements of the job. Therefore, there is a significant return on investment when a utility invests in its own employees.

Frequently, training and education are the first budget line to be reduced when there are budget cutbacks, yet they are some of the most important tools available to help a utility become more efficient. Decreasing or eliminating training and education only sets back the utility's workforce and causes the utility to fall behind as technology advances.

Capability development of a utility's workforce can be accomplished through robust training and education programs, providing continuing education funding support to employees, and establishing in-house job shadowing programs. Additionally, developing clearly defined career road maps that employees can use to understand how they can navigate future career paths in an organization will help employees identify and follow a possible future for growth.

Employee coaching is an incredibly effective way to grow and develop employees. However, to execute it, supervisors and managers must be trained on how to coach their employees. Many useful coaching programs are available to utility managers toward accomplishing this goal. For example, Gallup Poll uses a strength-based approach to coaching and developing employees. Coaching is also useful in training employees on the specific requirements and nuances of their area of responsibility. No two organizations are exactly the same, and each one has its own unique requirements that can only be passed down through coaching. Several resources and companies that specialize in staff development are available to help a utility establish a sustainable program.

Management and Leadership Development

More-specific employee development critical for a successful utility is in the area of management and leadership. There are several types of leadership positions in water utilities. They include but are not limited to crew leads, frontline supervisors, division/department managers, and executive directors. Significant numbers of people retiring from the utility industry are in leadership positions. A utility must respond to this issue through the development of its workforce, preparing them to move into these positions as they are vacated. This highlights the importance of coaching and mentoring programs to ensure that employees who are identified for future leadership positions are effectively prepared for the position when it becomes available.

As employees move from being individual contributors to positions requiring them to lead others in the organization, the focus increasingly must shift from technical intelligence and knowledge to emotional intelligence. Oftentimes utilities make the mistake of promoting the most technically proficient individual to supervisor because that person is so strong technically, even though the person has had no training in how to supervise and lead people. This type of promotion often results in situations where the new supervisor is ill equipped to lead and does not exhibit the inherent skills necessary to manage. Incorrect promotions can cause significant damage and turnover in an organization.

Several leadership development programs are available to utilities to assist in this area of workforce development. For example, the Rocky Mountain Section of AWWA has a supervisory certificate program that has successfully trained new supervisors and managers to prepare them for their new leadership duties. Additionally, several universities across the country offer two-week leadership programs that can prepare workers for leadership roles. Many of these types of leadership programs can be found through your local AWWA section.

EMPLOYEE RESOURCE POOLS

Several employee resource pools are available to utilities including graduates of high schools, colleges, and universities; veterans; and workers retrained from other industries.

High School Students

Utilities in Texas and Colorado have partnered with school districts to create programs promoting water careers to students. These programs provide students with skills and an immediate career path following high school. State regulatory curriculum is integrated with the schools' curriculum so that graduating seniors will have the necessary knowledge and classroom credit to sit for the entry-level operator license exam. These programs give students a working knowledge of water utilities and give them an idea of what a career in the water industry can provide. At the very least, these students will be informed customers. If a water utility does not have a program already, the utility is encouraged to contact its local school district to establish a local program. In addition to the curriculum and field experience, students also train and participate in Meter Madness Jr. competitions where they are timed to see who can fully assemble a water meter the fastest.

Colleges and Universities

Thanks to technology, social media, and the stiff competition for prospective students, college recruiting has become an art. Companies are enticing students with polished marketing materials, outstanding campus experiences, and visits. A strategically based college relations program looks at the challenge long term and is not just focused on filling open positions or short-term results. Organizations that experience the greatest success with their college relations program do not abandon the partnership when they are fully staffed. Instead they maintain their ties and target campus activities.

Depending on the skills and educational background that utilities are seeking, they may target four-year universities, master's programs, or two-year associate degree programs. While in the past many operators may have been hired directly with a high school education, hiring operators from two-year-associate degree programs may be beneficial for some utilities because of the increased complexity and advanced skills needed to operate many facilities. These two-year programs are focused on training water and wastewater operators and are taught at two-year and four-year colleges throughout the United States.

Many companies hire college interns from both two-year and four-year programs and convert them to full-time hires. The key to a successful college relations program is ensuring that interns are provided with challenging work that is valuable to the utility and is related to their majors. It also critical that they continue that focus for the duration of their internship.

Sending the right representative to sell the organization also is crucial to a successful program. Company representatives must be well-trained professionals who can answer questions and effectively represent the industry and the organization.

Prompt follow-up with students regarding the selection and hiring process is very important. Communicate honestly and frequently with students about what they can expect and when they can expect it.

Veterans

The water industry has identified veterans as an underutilized labor pool to meet its needs. Veterans bring many desirable values and abilities to the workplace. They have an accelerated learning curve because they are in possession of knowledge, transferable skills and abilities, and the ability to quickly acquire new skills to achieve success.

Veterans have leadership skills because the military trains them to lead using practical methods to manage staff and achieve their goals in a timely manner. The military stresses teamwork, giving veterans an understanding that genuine teamwork grows out of personal responsibility to the team, and they learn how best to motivate diverse groups to achieve mission goals. Veterans often have experience working alongside a diverse group of people with different ethnic, racial, and religious backgrounds, as well as different mental or physical capabilities.

The veteran applicant pools are trained to provide efficient performance under pressure. They are proficient in accomplishing tasks on time and in the face of adversity and stress thanks to an internal code of dedication and perseverance that they develop during their military service. Veterans are trained to triumph over adversity. They also have the ability to overcome challenges and obstacles through strength, determination, and a code of honor.

Veterans have respect for procedures. They value accountability and have an understanding of the importance that policies and procedures play in yielding stability, safety, and productivity. Veterans often have knowledge of international and technical trends pertinent to business. They also have integrity and are likely to have consistency of actions, values, methods, measures, principles, expectations, and outcomes.

All of the qualities just mentioned are very desirable to water utilities and make candidates who are veterans a great fit for water industry positions at all levels. Employers want to hire someone who either has or can get the skills and the training required to successfully accomplish the requirements of the position. Veterans often have the necessary experience. But they may need help in translating the skills and experience gained by military service into civilian job requirements. If they do need training, with an appropriate career plan federal agencies will often pay for training. Additionally, several programs provide financial incentives to employers hiring veterans.

Retraining Employees From Other Industries

Water utilities have the opportunity to recruit qualified employees with extensive and diverse experience in areas such as environmental sciences, human resources, information technology, lab operations, engineering, electrical or industrial maintenance, construction, accounting, and customer service. Hiring an employee with a foundational knowledge and work experience in another industry can significantly reduce the staffing challenges experienced when the utility's recruitment efforts are focused solely within the water industry. There are a number of career opportunities in the water industry in which the required skills and competencies may be obtained from various industries (manufacturing and gas and electric utilities) and/or educational institutions.

Local technical colleges offer certificate programs in the following areas:

- Industrial electrician
- Industrial construction
- Certified construction worker
- Industrial systems technology
- Water and wastewater technology

These technical programs may offer training, retraining, and certification in the areas of basic electrical maintenance skills; industrial safety; industrial wiring; several areas of industrial maintenance including electronics, programmable logic controllers (PLCs), and instrumentation; and mechanical maintenance on pumps, piping, and chemical feed systems.

Additionally, state and local labor organizations may offer excellent resources in the areas of job training, retraining, certification testing, and skill assessments. The US Department of Labor's (DOL) Employment and Training Administration (ETA) funds job training programs to improve the employment prospects of adults, youth, and dislocated workers. Programs are aimed at improving employability of job seekers and include adult services, apprenticeship, dislocated workers' programs, veterans' programs, and others.

The DOL adult services programs assist adult workers wishing to change careers with workforce preparation and career development services and help employers recruit and hire skilled workers. Programs provide high-quality employment and training services that address the needs of individuals in the areas of training, retraining, and skills upgrade. Additionally, investments in adult services are targeted to move workers into postsecondary educational pipelines and career pathways to prepare more workers to enter into and advance in good jobs in the high-growth and emerging occupations of the global economy.

KNOWLEDGE MANAGEMENT AND SUCCESSION PLANNING

Knowledge management (KM) is rapidly emerging as a major issue for water utilities as the workforce is aging and significant numbers of key employees are nearing retirement age. Workforce issues that are driving the need for improved KM include

- Baby-boomer retirements and the consequent loss of latent knowledge
- Increasing health and safety, environmental, and water quality regulations

- Changing facilities, technologies, and equipment
- Competition with other industries for skilled employees
- Increasing public expectations with regard to customer service, efficiency, and cost
- Industrywide underinvestment in the processes required to support staff preparedness (such as documentation, training, mentoring, and use of knowledge management systems)

Knowledge can be explicit or tacit. Explicit knowledge is knowledge that is written down, codified, and stored. Examples of explicit knowledge include reports, documents, diagrams, and web content. The challenges with explicit knowledge are keeping it up to date and making sure that it can be found and accessed when needed by employees.

Tacit knowledge is the knowledge that resides within the employee and is generally not written down. Tacit knowledge also is referred to as *know-how knowledge, experience,* or *skill.* A significant portion of operating information is tacit knowledge, which is in danger of being lost when the employee leaves the utility. For existing employees, the challenge is to keep tacit operational knowledge up to date, ensure that key operational knowledge is captured, and make it available when it is needed.

What Is Knowledge Management?

The goal of knowledge management is to ensure that the right people have the right skills at the right time to perform work needed efficiently and effectively.

Knowledge management focuses on identifying key organizational knowledge and capturing it in a way that can be maintained, retrieved, updated, and effectively disseminated to the people who need it.

Knowledge management is a combination of people, processes, and technologies. The people aspect focuses on an organization's culture (sharing of information, collaboration, and communication). Processes include how knowledge is captured, stored, maintained, and disseminated. Technologies include the selection and implementation of appropriate business support and communication systems that are aligned with the people and processes. Successful KM programs combine these three elements. In many cases, significant improvements in knowledge management can occur with initiatives focused on low-cost options that target people and processes first before making large investments in supporting technologies.

The following chapters discuss how to get started with knowledge management, building momentum and sticking with it, common knowledge management tools, and case studies of existing KM programs.

Getting Started

It is critical to define strategies that will optimize return on investment because the funding and staff time available to utilities for knowledge management programs are limited. A few basic questions should be asked before getting started:

- What is the level of knowledge gap in various business and operational areas?
- What is the knowledge we need to capture and what should be the priority of capture?
- What are the consequences if this knowledge is not where it needs to be?
- How should the desired knowledge flow in our organization—from whom and to whom?
- What are the best pathways for this knowledge to flow?

Answering these questions first will allow you to better decide which tools, strategies, and technologies will help you to capture the desired knowledge (Edge 2010).

Suggested guidelines for development of a program are summarized in Table 4-1.

NEEDS ASSESSMENT

It is important to identify the unique issues and needs of your organization. Resist moving into problem-solving mode at the start. Instead, take the time necessary to identify key needs and develop a workable targeted strategy. This assessment should combine the insights of senior management with those of staff from other levels in the organization. The following sections outline the components of the needs assessment.

Identify Mission-Critical Functions and Job Categories

Every employee in an organization has an important role to play. However, in a job climate characterized by high turnover and change, it is not feasible (either in terms of cost or staff time) to make an equal investment in every job category and work process. Therefore, it is useful to conduct a workforce development needs assessment to identify the core operational responsibilities of the organization (e.g., water supply, water quality, and environmental stewardship) and the job categories where deficiencies in staff knowledge or skills could put core functions at most risk. This type of analysis helps an organization set priorities based on documented business risks.

Table 4-1 Suggested guidelines for getting started with a KM program

Program Component	Suggestions
A. Needs assessment	1. Identify mission-critical functions and job categories.
	2. Assess the culture of your organization.
	3. Assess the effectiveness of the processes used by your organization to ensure that staff in mission-critical job categories have (or have access to) the information and skills needed to perform quality work.
B. Implementation	1. Find an approach and suite of staff preparedness tools that fit the needs, culture, and priorities of the organization.
	2. Focus on how employees will access and use information so that your delivery systems will reflect the need.
	3. Adopt a phased approach.
	4. Identify staff who will take the lead in the knowledge management process.
	5. Look for and create learning opportunities.
	6. Look for opportunities to optimize the cost-effectiveness of your efforts by partnering with others.
	7. Create a culture that encourages information-sharing and collaboration.
C. Sticking with it	1. Recognize that development and maintenance of staff knowledge are a long-term commitment to changing how you do business.
	2. Reinforce the successes of the organization in knowledge management by acknowledging achievements and mentoring others in the industry when knowledge management efforts prove effective.

Assess the Culture of Your Organization

An organization's culture is an organization's way, identity, and pattern of dynamic relationships—its reality. It has everything to do with implementation and how success is actually achieved. A management concept such as knowledge management, no matter how good, will not work in practice if it does not fit the culture of the organization. An organization can have the most superb strategy, but if its culture is not aligned with and supportive of that strategy, the strategy will either stall or fail. Culture establishes and underpins order, structure, communication patterns, expectations and priorities, decision-making practices, management practices, and other management approaches and systems. It is important to understand the organization's culture and either implement knowledge management in a way that aligns with that culture or implement a cultural change program prior to initiating knowledge management.

Assess the Effectiveness of Your Processes

Organizations vary in the investments they make in documentation, staff training, and knowledge management. Often utilities have defined in detail the qualifications and credentials needed by candidates to obtain a job but have not defined the utility-specific knowledge and skills needed to perform the work successfully (e.g., the technologies, equipment, and procedures associated with the utility's water treatment and distribution processes). Similarly, utilities often have no process for identifying the gap between what employees need to know in order to perform the work and what they actually do know. Therefore, during the needs assessment phase it is critical to perform a gap analysis to assess gaps, if any, between the current state and the desired state with respect to existing tools and processes that support knowledge capture, information update, and dissemination.

IMPLEMENTATION OF KNOWLEDGE MANAGEMENT

Approach

It is critical to find an approach workforce management that fits the needs, culture, and priorities of the organization. Although most organizations in the water—wastewater industry face similar problems such as staff turnover and changing requirements, they have adopted diverse means to address them.

Accessing and Using Information

It is important to focus on how employees will access and use information so that your delivery systems will reflect the need. The purpose of knowledge management in the water industry is to meet the information needs of both current and future employees. Both current and future needs should be taken into account. However, a priority should be placed on making critical information easily available to those who need it, when they need it, in a form that is usable to them.

Employees learn in different ways and may need to access information from a variety of settings. Therefore, consider multiple packaging of important messages. Key information might be reflected in a written procedure, a video, a PowerPoint presentation, workflow diagram, interactive online training tool, field guide, or a scenario-based training session. For critical tasks, it may be worthwhile to prioritize multiple packaging of key messages over having a hard-copy write-up on all minor procedures.

Phasing

Consider adopting a phased approach. It is rarely feasible to address all of the knowledge management needs of the organization simultaneously. Some activities may be more appropriate for certain parts of the organization. For example, certain competency assessments may be appropriate for some workgroups but not others. Some activities may be appropriate for all parts of the organization, such as using information technology to support knowledge management. Effective implementation may require an incremental approach of selecting, storing, and accessing knowledge.

It can be very helpful to conduct a pilot study on one or two selected knowledge management processes and one or two functional areas. A pilot study is essentially an initial investigation used to test the feasibility of ideas and procedures that you could effectively use utilitywide as part of your larger KM plan.

Knowledge Management Leaders

Identify staff who will take the lead in the knowledge management process. It is valuable to have one or more knowledge management champions who are willing and able to lead knowledge management programs. Key staff groups can work as knowledge teams, developing methods and taking ownership of the process. The KM leaders do not have to come from the ranks of upper management. In fact, in some organizations it is best that they do not. But they do need to possess good communication skills, be respected in the organization, and able to see the big picture and how different components fit together. They will also need a persevering attitude that will keep the organization on track through the inevitable challenges that lie ahead.

Learning Opportunities

Utilities need to look for and create learning opportunities. Your staff members can learn from each other and from your organization's retirees. They also can learn from staff at other utilities and through a combination of formal and informal coaching and mentoring. Training can be provided by your own personnel, using online and video resources, and through external sources. Learning should be tracked to understand gaps, and use past training for learning opportunities.

Cost-Effectiveness

Implementing workforce management improvements can lead to opportunities to optimize the cost-effectiveness. Most utilities have had insufficient resources to provide a full range of documentation, training, and knowledge management programs, even in better economic times. In a time of diminished resources, many large organizations may be inadequately prepared, and small ones may be even more challenged. Historically, most organizations in the water and wastewater industry have independently addressed the training and knowledge management needs of their staff, with some assistance from professional associations and community colleges. The efficiencies of scale potentially possible through increased coordination have yet to be realized. Video-conferencing, shared development and use of training materials, use of shared knowledge management systems, and shared scheduling of training opportunities are examples of areas where utilities could both leverage their investments and increase staff learning.

Performance Measurement

Utilities will need to measure how well the tools and strategies are working. Begin with the end in mind—establish your desired outcomes and then work backward to design KM activities and measures that focus on those outcomes.

Many helpful measurement frameworks are available. One example is the American Productivity and Quality Center (APQC) framework, a measurement framework showing the relationship of inputs, process, and outcomes. For example, inputs include investments of time, resources, and technology costs. Process includes participation levels and the time it takes to transfer and document knowledge. Outcomes can be the objectives of your organization, which could include specific knowledge capture and transfer, cross training, and/or employee retention.

STICK WITH IT

It is important to recognize that development and maintenance of an institutional KM program is a long-term commitment to changing how you do business. Documentation, staff training plans and materials, and knowledge management systems must be embedded in work processes so that knowledge management becomes a sustained, ongoing effort. Timely updating of materials is necessary for them to remain relevant and beneficial. One-time investments of staff time and funding are not enough. Investment in the knowledge infrastructure of the organization will be crowded out by daily operations and maintenance responsibilities unless money and staff time are dedicated for this purpose.

Utilities must reinforce the successes of their knowledge management program by acknowledging achievements and mentoring others in the industry when knowledge management efforts prove effective.

REFERENCES

American Water Works Association (AWWA). *AWWA State of the Industry Report*, 2006, 2007, 2008, 2009, 2010, 2011, 2012, 2013, 2014, 2015, and 2016. Denver, CO: AWWA.

AWWA Knowledge Management Subcommittee of the AWWA Workforce Strategies Committee. Committee Report: Knowledge Management for Addressing Workforce Issues. 2001. *Jour. AWWA*, 103(8):44–50.

AWWA Research Foundation. 2003. *Workforce Planning for Successful Organizational Change.* Denver, CO: AWWA Research Foundation and AWWA.

AWWA and Water Environment Federation (WEF). 2013. Recruiting and Training Veterans for Careers in the Water Sector. Posted on the Work for Water web site sponsored by AWWA and WEF.

AWWA State of the Industry Report, 2006, 2007, 2008, 2009, 2010, 2011. 2012, 2013, 2014, 2015 and 2016. Denver, CO: AWWA.

Baywork, Bay Area Water/Wastewater Workforce Development. http://baywork.org/

Brueck, T., M. Isbell, D. O'Berry, and P. Brink. 2010. *Water Sector Workforce Sustainability Initiative.* Denver, CO: Water Research Foundation.

Edge, K.M. 2010. Need title and publisher.

Fletcher, M.A. 2011. Veterans' Unemployment Outpaces Civilian Rate. *The Washington Post*, Oct. 16, 2011.

Goodburn, C., Farmer, G.; Rousey, M., "The Littleton/Englewood Wastewater Treatment Plant Knowledge Management Initiative: Using an Online Manual to Capture Knowledge." 2010. Cindy Goodburn, Littleton/Englewood Wastewater Treatment Plant, CO, Greg Farmer, Littleton/Englewood Wastewater Treatment Plant, Mike Rousey, Brown and Caldwell

Jack, J.R., MHS, CRC, Employment Coordinator, Department of Veterans Affairs. 2012. Vocational Rehabilitation and Employment (VR&E) VetSuccess Program. Presented as a Webinar by the Rural Water Association. March 2012.

Mann, J., and J. Runger. 2010. State of the Industry Report 2010: How Water Professionals Are Meeting Ongoing Challenges and Economic Uncertainty. *Jour. AWWA*, 102(10):44.

Mann, J., and J. Runger. 2009. State of the Industry Report 2009: How Tough Times May Point the Industry Toward a More Sustainable Future. *Jour. AWWA*, 101(10):36.

Mann, J., and J. Runger. 2008. State of the Industry Report 2008: Charting the Course Ahead. *Jour. AWWA*, 100(10):61.

Mann, J., and J. Runger. 2007. State of the Industry Report: An Eye to the Future Ahead. *Jour. AWWA*, 99(10):58.

Manning, A. 2008. *Workforce Planning for Water Utilities—Successful Recruiting, Training, and Retaining of Operators and Engineers to Meet Future Challenges*. Denver, CO: AWWA Research Foundation and US Environmental Protection Agency.

McCain, K. 2012. Veterans An Untapped Resource to Mitigate the Upcoming Workforce Crisis. *Jour. AWWA*, 104(8):49–54.

McPherson, J. 2010. Union Sanitary District. Competency Based Training System Model.

Olstein, M.A., D.L. Marden, J. Voeller, J.D. Jennings, and P.M Hannan. 2005. *Succession Planning for a Vital Workforce in the Information Age*. Denver, CO: AWWA Research Foundation and Water Environment Research Foundation.

Stockdale, C.B. 2012. 10 Jobs Americans Can't Live Without. *Reader's Digest* Online. www.rd.com/sslideshows/10-jobs americans-cant live without/.

US Bureau of Labor Statistics. 2011. Unemployment Rates. www.bls.gov.

Vogel, S. 2012. Obama to Announce Veterans Job Corps, *The Washington Post*, Feb. 3.

Water Research Foundation. 2012. Competency Model Development and Application to Meet Water Utility Workforce Needs. Denver, CO: WRF.

Useful Web Sites

Baywork, Bay Area Water/Wastewater Workforce Development, http://baywork.org/

http://baywork.org/wp-content/uploads/2013/02/HOWTO-Create-a-High-School-Water-Training-Program.pdf

http://baywork.org/resource/how-to-city-of-waco-create-a-high-school-water-training-program/

http://www.bls.gov/ooh/production/water-and-wastewater-treatment-plant-and-system-operators.htm#tab-1

Columbus Technical College, Columbus, Georgia http://www.columbustech.edu/

Cuyamaca College, El Cajon, CA/ http://www.cuyamaca.edu/wwtr/Degrees.asp

http://www.careeronestop.org/

Georgia Association of Water Professionals/ http://www.gawp.org/

http://getintowater.org

http://h2opportunity.net

United States Department of Labor/ http://www.dol.gov/

http://www.rd.com/slideshows/10-jobs-americans-cant-live-without#slideshow=slide11

http://watercareers.org/

http://workforwater.org

http://workinwater.org

This page intentionally blank.

Chapter 5

Information Management: Whoever Manages Information the Best Wins!

This chapter addresses the fundamental changes information management enables within water–wastewater utility operations. The purpose of this chapter is to provide the reader with an overview of the key information management issues and, where applicable, provide references for more in-depth explanation or clarification.

More data are available than ever before, assisting managers to more closely monitor and better manage their systems. Here you will find the key information management trends occurring within the water industry, an outline of their importance to the manager, and suggestions to make the most of these new capabilities. Some key observations include:

- Information technology (IT) is changing faster than any other field. Studies show that IT professionals have the highest "educational needs stream" of any other occupation, making it challenging and almost impossible to stay current with and knowledgeable of emerging technologies.
- As the availability, quantity, and significance of data grow within the utility, so does the importance of sharing that data among multiple stakeholders. The effective utility will be one that embraces collaboration and seeks to develop and maintain an integrated enterprise management framework.

- Information technology is considered a fundamental tool in the operations and success of a utility. IT professionals touch on almost every facet of the organization and should be involved in strategy development and the planning process.

DATA FLOWS

Source to Consumption

Countless data points are being generated every minute at a typical water utility that are monitored and consumed within the utility's core business systems.

Core utility information systems that are involved are geographic information system (GIS), supervisory control and data acquisition/distributed control system (SCADA/DCS), hydraulic modeling, laboratory information management system (LIMS), computerized maintenance management system/enterprise asset management (CMMS/EAM), capital improvement plan (CIP), customer information system (CIS), meter reading—automated meter reading/advanced metering infrastructure (AMR/AMI), enterprise resource planning (ERP), mobile workforce, and stakeholder engagement (web services).

GIS (geographic information system). Integrates hardware, software, and data for capturing, managing, analyzing, and publishing all forms of geographically referenced information. As much as 80 percent to 90 percent of information used by water utilities is georeferenced. GIS locates utility network assets like mains and manholes, devices like valves and pumps, and meters. It also stores data about each asset, including manufacturer, pipe type, customer complaints, repair history, and water quality, to name a few.

DCS/SCADA (distributed control system/supervisory control and data acquisition). Gathering and analyzing real-time or near real-time sensor data used for monitoring the treatment plant and transmission and distribution processes. Key data contained within the DCS or SCADA system include flows, levels, pressures, chemical use, energy use, filter backwash, and so on.

Hydraulic modeling. A simulation tool that illustrates the effects of changing demand and conditions on water distribution and wastewater collection systems. Models simulate and analyze the performance of a network using what-if scenarios. Key data needed for the hydraulic model include physical system information, precipitation data, pump curves, well and booster pump operational controls, and hourly water usage patterns. Models often integrate with GIS because it is usually the utility's most up-to-date data for model input and because of its visualization capabilities.

LIMS/LIS (laboratory information management system/laboratory information system). Used for collecting and monitoring water quality lab samples. Key data within the LIMS include laboratory results for regulatory testing. Needs to include field data collection and real-time sensor data. Also integrates with enterprise resource planning (ERP) and SCADA.

CMMS/EAM (computerized maintenance management system/enterprise asset management). CMMS and EAM (often used interchangeably) are used for planning, scheduling, monitoring, and analyzing maintenance activities, including corrective (reactive) and preventive (proactive) maintenance. Some key data within the CMMS include pipe size, material, age, condition, manufacturer, customer complaints, condition rating, and maintenance history.

CIS (customer information system). Used for customer account management, meter reading processing, billing, collecting, and service order processing. Integration: GIS, EAM, meter data management, CMMS, and mobile workforce management.

Support and ancillary systems. Other information systems are also typically needed to support the core business applications. These typically include capital improvement plan (CIP), enterprise resource planning (ERP), stakeholder engagement, and other systems.

From the Field

Work performed in the field is as broad as the utility itself, and the complexity of the work can range from a simple meter reading verification (sometimes called a *check read*) to complex processes such as a water treatment plant upgrade project. Here, however, we will focus on routine maintenance work—everyday activities performed in support of normal utility operations. Work requiring actions to be performed in the field can originate from any department or position within or even outside of a utility. However, a majority will come from key departments such as finance/billing, engineering, customer service, and operations and maintenance. Table 5-1 contains an example of a few types of work generated by each department.

Additionally, some work will be generated in the field. For example, a simple investigation field work order issued by the finance/billing department could result in multiple additional work orders, such as meter exchanges or repair activities. Work completed, whether initiated from the office or in the field, does require that the results be reported to the department or system that originated the original request, but this same data can be automatically distributed to other applications. More useful data equals better understanding and lower risk. This data sharing also provides a more up-to-date and consistent picture of the utility.

The Intelligent Water Utility

Historically, utility systems have included desktop software that provided functionality and output specific to one group or department. This software usually has fallen into one of the following divisions or business units: planning and engineering, operations and maintenance, finance and administration, or infrastructure management and construction. Some of the software systems used to support these departments include Microsoft Excel for management; hydraulic modeling for engineering; supervisory control and data acquisition (SCADA) for operations; computerized maintenance management system (CMMS) for maintenance; customer information system (CIS); automated meter reading/advanced meter infrastructure (AMR/AMI) for customer service or finance; and geographic information system (GIS) to support utilitywide mapping. When integrated, these and other systems provide an enterprisewide management platform. Using information from multiple systems, bringing complementary perspectives, and adding value to work and effort are required to fully understand, predict, and justify return on investment (ROI) and resiliency within a utility.

Table 5-1 Examples of field work generated, by department

Finance/Billing	Engineering	Maintenance
Investigation	Water main break/leak	Valve exercising
Meter exchange	Line shutdown	Hydrant flushing
Meter installation	Check tank levels	Large meter maintenance
Check read		

With the growth and prevalence of GIS, utilities are evolving beyond maps, and moving into standardizing data and data sets allowing differing systems to reside in a common database, for visualization, analysis, and mapping (see Figure 5-1 as an example). We are also seeing software and data migrating to the web, with cloud hosting enabling access to information from any device (smartphones, tablets, etc.) by any utility worker at a lower cost of ownership and ready to use. These GIS web services share data across the utility, in real time, providing a lightweight, configurable operating platform for business processes management, knowledge-based decision support, and mobile workforce.

Figure 5-2 shows an example using GIS and work order data to provide a heat map of water main break and repair activity, helping the utility manager to determine areas of focus.

Starting at the Meter

The utility's revenue stream starts with the water meter. Meters are widely used as an equitable means of allocating the utility's costs to generate and deliver clean water across a wide range of consumers. Meter readings for bill generation are generally only captured on a quarterly, bimonthly, or monthly basis. Various levels of automation are used throughout the industry to capture water usage information, starting with manual readings where a meter reader walks a service area, visually inspecting meters and reporting into a route book, to advanced metering infrastructure (AMI) that has completely automated the meter reading process. AMI is an architecture for automated communication between a smart meter and the utility, providing real-time consumption data without leaving the office. The large amount of data (Big Data) that are collected has to be converted into a format that makes sense and that is usable across the enterprise.

However these readings are collected, the meter readings are brought back to the utility where the reading from the previous billing period is subtracted from the most recent reading—providing the utility with the amount consumed. This consumption value is converted by the utility's billing system into a dollar value based on the cost-per-unit volume. Other fixed charges are applied to cover the capital invested to deliver water

Figure 5-1 IDModeling: A screenshot of Sedarū command center

Source: Copyright 2013, IDModeling Inc.

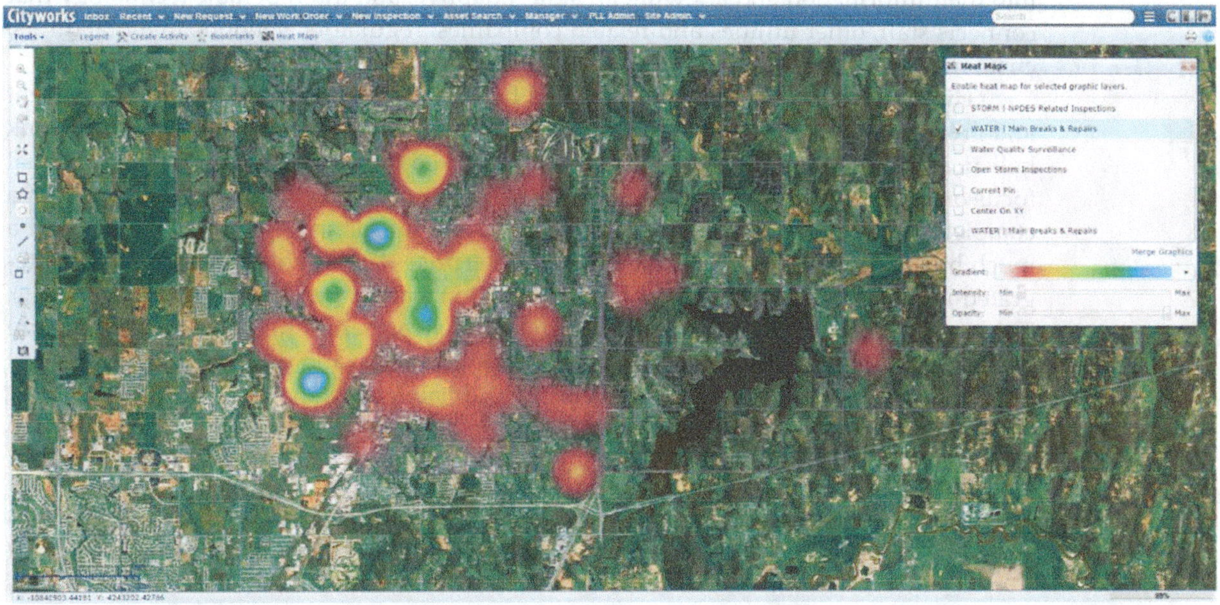

Figure 5-2 An example heat map of water main break and repair activity

Source: Azteca Systems Inc., Cityworks®, Copyright 2013.

including the cost of the meter, the service line, and other nonvariable costs. A bill is then issued. Although the paper bill remains the most common method, bills issued using electronic means are becoming more readily available and accepted. Electronic billing allows consumers to view their bills and make the payment online.

The utility's customer service department supports the billing process by addressing customer complaints and questions. By using accurate and timely information about a consumer's water use, billing and call history, and service details, the customer service representative (CSR) can be more knowledgeable and responsive to the consumer, enhancing the customer experience while achieving the appropriate level of service (LOS).

Also, if a customer's call results in the need to send a field crew to investigate or respond, the CSR through the CIS can issue a work order and the appropriate crew can be scheduled and dispatched to the site.

YOU CANNOT MANAGE WHAT YOU DO NOT MEASURE—DATA MANAGEMENT AND THE BUSINESS PROCESS

GIS as a Platform

A geographic information system (GIS) platform allows utilities to store, manage, and map spatial information that can be accessed anywhere, anytime, and on any device. The result is that enterprise systems and information can be connected and leveraged in real time, increasing collaboration and efficiency across the organization. The power of GIS has always come from its emphasis on collaboration, integration, and communication. The difference now is that, with GIS as a platform, the scale of collaboration and integration has increased, enabling much broader and more direct communication and interaction. GIS as a platform connects and leverages all enterprise systems in real time. Instead of having data in separate databases in different departments, utilities can have all their data (apps, imagery, maps, social media, etc.) in one graphic expression, providing a window into their information and allowing an agile, adaptive, and interoperable technology that connects staff, other utilities and related companies, regulators, vendors, manufacturers, customers, and the public.

Historically, water has been measured by quantity and quality. Because of current and predicted water stresses, we need to include adaptability and sustainability, and look holistically at the water cycle. Utilities are being asked to shift from a narrow operational focus on water service delivery and wastewater treatment and into the management of a complete watershed. GIS offers a common platform for managing water resources on a much broader scale. Using GIS, all types of enterprise data related to water can be integrated into a single real-time dynamic information system (see Figure 5-3, for example). This approach cuts across traditional municipal delivery areas to include infrastructure design and operations, land-use planning, public education and participation, emergency planning and response, pollution prevention, and habitat and shoreline restoration.

Computerized Maintenance Management Systems

As technology has advanced, maintenance management has evolved from manual processes to computerized maintenance management systems (CMMSs) or enterprise asset

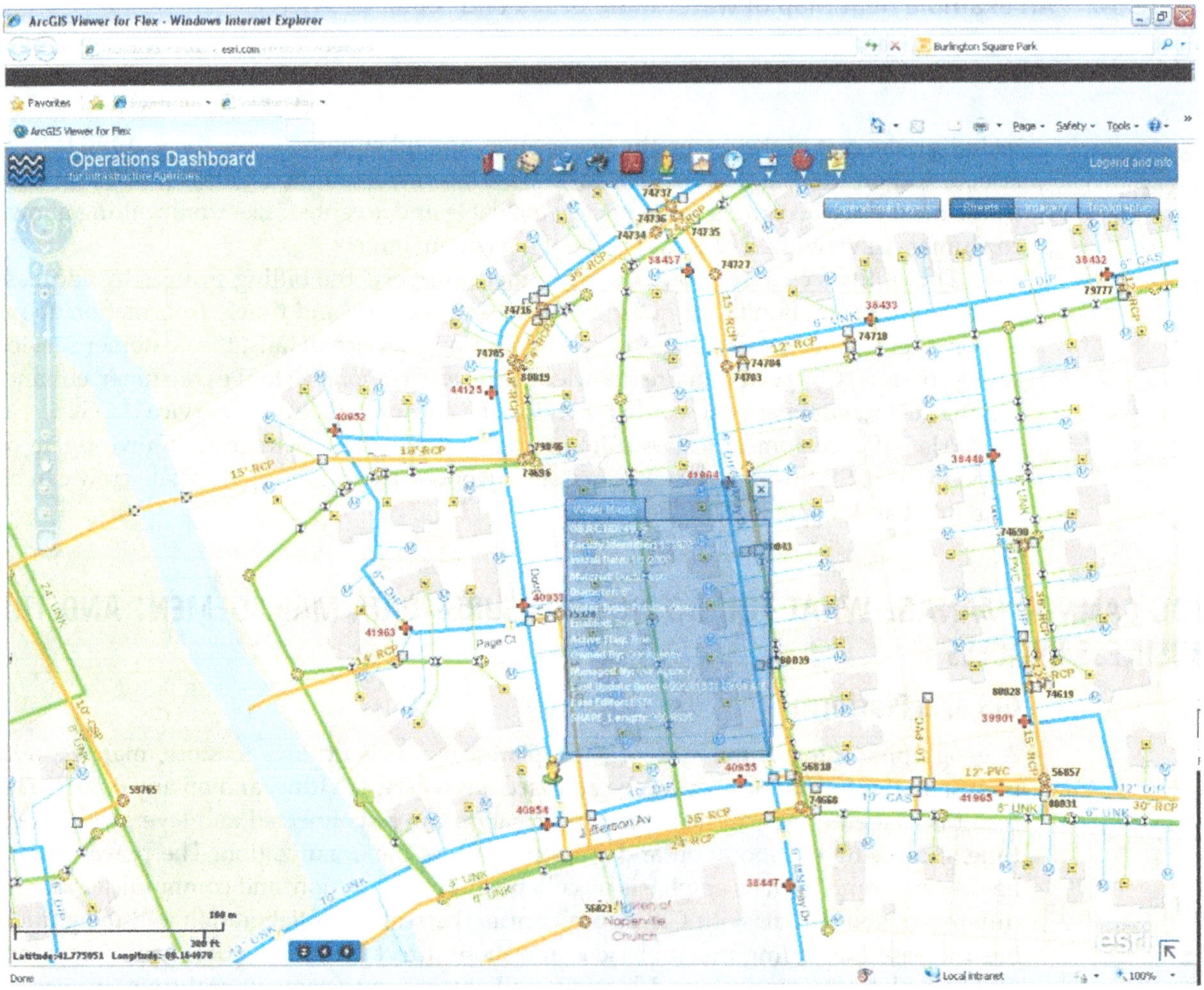

Figure 5-3 Screenshot of ArcGIS viewer

Source: Esri, Copyright 2013.

management (EAM). A CMMS helps utilities manage information about their maintenance and operations in the office and in the field. Modern GIS-centric systems allow users to interact in a map-driven environment and operate in a more visual way. By using technology to automate and document workflows, an organization can be more efficient, save money, and make better knowledge-based decisions.

A utility thinking about implementing or updating a CMMS should consider the following:

- Choose a CMMS system that fits the needs of the utility: size, operations, and key business processes and workflows
- Identify a team of champions and project owners for redundancy and accountability
- Involve users and stakeholders from the beginning
- Maintain the software and staff to administer the system
- Keep skills current and up to date

Maintenance management is a key part of any operation, and CMMS software enables maintenance professionals to work smarter by streamlining workflows and generating usable data that have a direct effect on an organization's bottom line.

AMR/AMI/Smart Networks

Big Data are a growing trend in the IT business and AMI/Smart Networks are great examples. The meter-reading process generates a huge amount of real-time data.

Advanced metering infrastructure (AMI) enables readings to be gathered without the need to send a meter reader into the field. Data collectors installed on water towers or utility poles can now capture the radio frequency signals from the water meters and deliver the data to the utility office. Figure 5-4 shows the evolution of data gathering methods. Although each of these functions differs significantly in its use of technology, they all have the primary function of gathering data to generate a water bill.

The current trend is to go beyond meter reading and use these communications networks to gather data from a variety of inputs. Smart Networks include devices such as

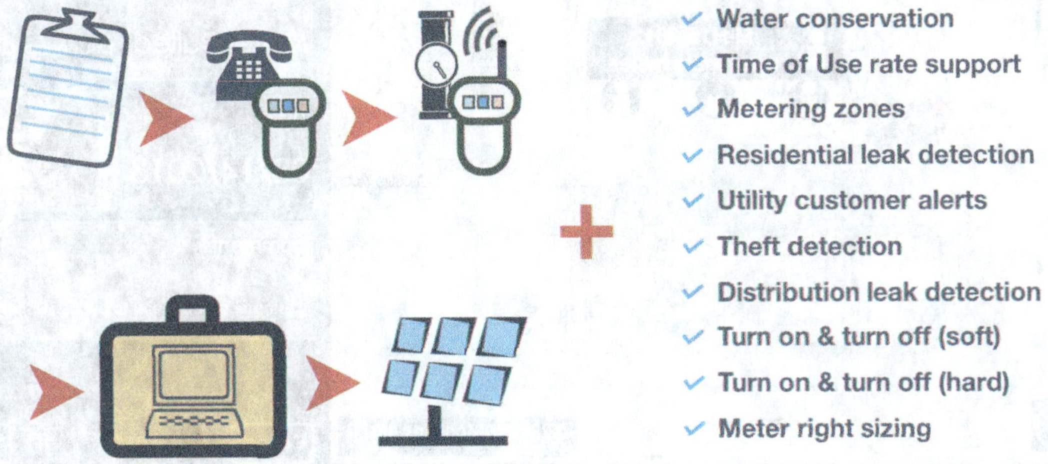

Figure 5-4 Evolving data gathering

Source: Neptune Technology Group Inc., Copyright 2011.

acoustic monitors to identify leaks in the distribution mains; and pressure, temperature, pH, and other sensors to provide real-time information on the factors impacting water quality. Some systems also extend from basic monitoring to include some elements of system control. For example, pressure-regulating devices will enable the automatic reduction of nighttime water pressure in the distribution system to reduce the stress on pipes and their resulting leaks.

Meter data are now being used by other critical systems providing a more accurate view of current utility operations (as shown in Figure 5-5) and a better understanding for future planning. Systems have been developed to enable meter data connectivity, allowing the data to be accessed and applied across the utility for workflow automation, improved decision-making, and improved customer response.

SCADA

Supervisory control and data acquisition (SCADA) is a mission-critical system that monitors and controls the utility's water treatment plant and transmission and distribution processes. A utility's SCADA system consists of the following subsystems:

- A human-machine interface (HMI) presents processed data to the utility operator to monitor and control the various treatment and pumping processes

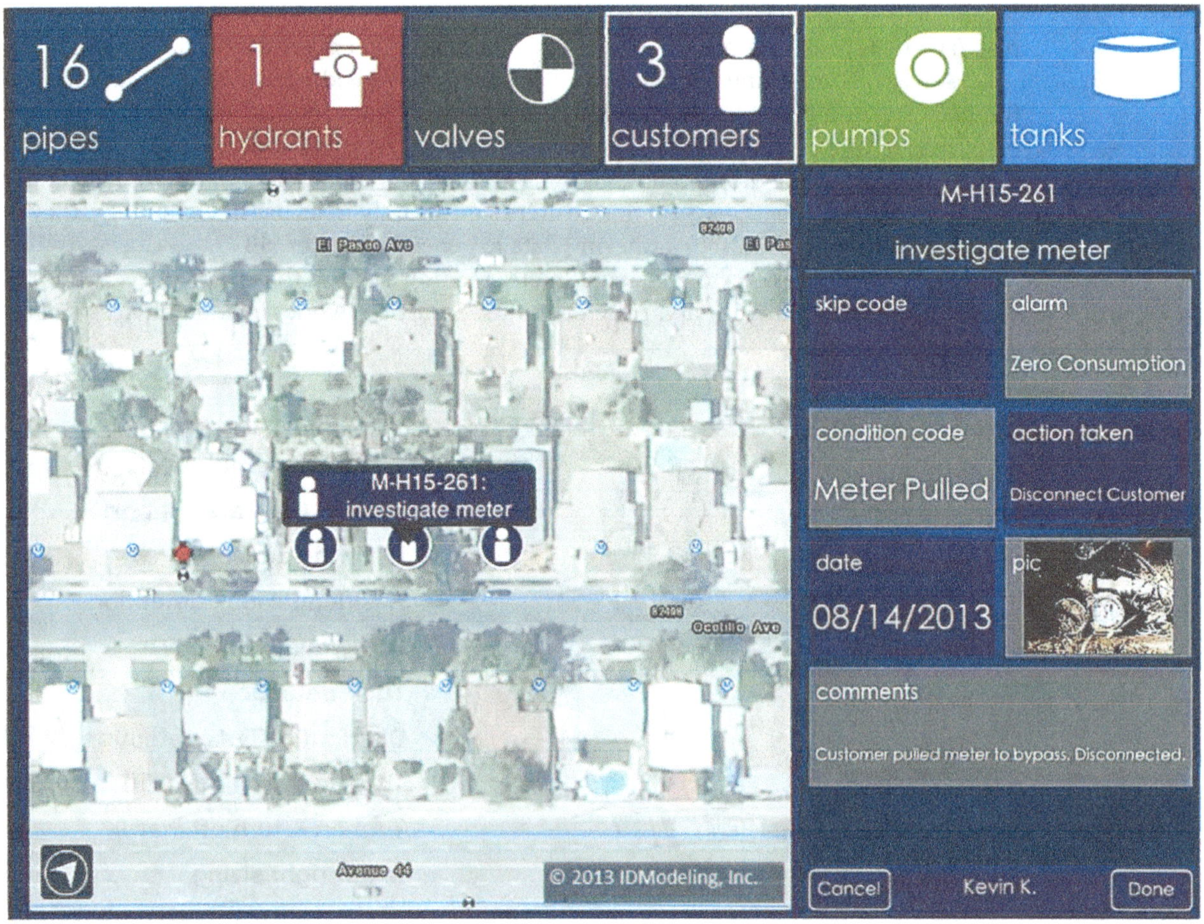

Figure 5-5 Meter data connectivity; commercial product example

Source: IDModeling, Copyright 2013.

- Remote terminal units (RTUs) and programmable logic controllers (PLCs) are field devices that connect the sensors in the process, convert the sensor signals to digital data, and send the digital data to the SCADA system
- A communication or telecom infrastructure connects the SCADA system to the field devices
- Various process and analytical instrumentation complete the system

Today's SCADA systems are utilized to do much more than just provide process control. Modern SCADA systems can collect data for regulatory reports, capture operational data for planning and decision making, optimize energy and chemical use, validate hydraulic models, help in identifying water line loss, and attain sophisticated asset and maintenance management functionality using real-time data.

Effective utilization of a modern SCADA system requires a comprehensive plan for all aspects of automated control, including technology, networks and communication, security, governance, data quality standards, and operation. A SCADA master plan should be developed and executed to

- Establish and document the vision, strategies, and goals for effective systemwide operations control
- Identify and prioritize business, operational, and technical requirements
- Establish a coordinated, prioritized program to meet operational goals
- Define specific short-term and long-term projects including costs, resource commitment, schedules, and dependencies
- Define and identify policies, procedures, organization, technologies, and a controlled means of managing change
- Establish buy-in among stakeholders and executive sponsors

Accessing operations data has not always been a user-friendly process. Data collection was often unreliable, and integration required costly customization. The result was that supporting and maintaining operational data were time consuming, and direct access to the data was limited to the SCADA or IT staff. Installation of an operations data management system (ODMS) helps in data capture through the use of a time-series database to provide a uniform view of a utility's operations data, regardless of the source. Time-series databases are commonly called *historians* and may be custom built, supplied by a utility's HMI vendor, or developed by an independent ODMS solution provider. An ODMS can help a utility manager

- Streamline creation and distribution of regulatory compliance and production reports
- Analyze incidents to determine cause and effect for establishing corrective actions
- Calibrate hydraulic models used for emergency response, capital planning, what-if analysis, and optimization of energy, chemicals, water loss, water conservation, and water quality
- Trend equipment performance to support condition-based predictive maintenance and asset renewal
- Record metering data to support bulk water delivery invoicing
- Monitor and track performance that supports operators' decisions

The success or failure of the SCADA system or any SCADA component can profoundly affect the utility's ability to maintain desired service levels.

Hydraulic Modeling

The hydraulic model is a decision-support tool used to simulate and predict system operating conditions. The model is used to locate and size water or wastewater system infrastructure that is required to support planned growth, but it is increasingly being used as a simulation operational tool to achieve energy management, water quality regulatory compliance, field maintenance prioritizations, and water loss reduction. Utility engineering develops the hydraulic model to reasonably simulate system conditions within a predefined set of known operating parameters. This process is called *calibration* and requires information from the distribution or collection system (pipes, hydrants, valves, manholes, etc.) in addition to facility data (pumps and tanks, for example) to calibrate against. Operating conditions are calibrated to include pressures, flows, tank levels, and water quality within a water system, and depth of flow or overflows within a wastewater collection system. Once the hydraulic model is reasonably simulating prescribed operating conditions, the utility engineer considers this model to be calibrated and have an acceptable level of confidence for planning and operational decision support (AWWA 2012).

The model is developed from information across utility systems, including GIS, CIS, and SCADA, supported by field maintenance. These system data are then supplemented by the institutional knowledge within the utility and engineering judgment guided by industry best practices to make practical adjustments to model parameters to improve calibration and increase the level of confidence around hydraulic model results. It is worth noting that (1) information collected across utility systems to build and develop the hydraulic model is maintained in a disconnected state and serves as a one-time data extraction that is quickly, if not instantly, outdated, (2) the calibration process is typically limited by utility resources, project budget, and schedule, and therefore represents an incomplete sample of the utility service area (AWWA 2012), and (3) the personnel who have utility institutional knowledge that grounds the model in practicality and usefulness are retiring at an alarming rate most utilities.

Global climate change, technology, and economic pressure are all global trends that are driving the call for real-time hydraulic modeling.

> Climate change increases the risk of volatility in water supply, thus making it impossible for the industry to rely on historical data for future planning. The main challenge for utilities is to reduce the effect of volatility on resources, infrastructure, and demand. We need flexible and smart systems that monitor, regulate, and anticipate changes in circumstances. (Deloitte Touche Energy and Resources 2012)

By using up-to-date data related to climate, infrastructure, and system performance, we can model volatile systems with volatility with a higher level of confidence. As a result, we can create simulation tools, such as the real-time hydraulic model, that are more appropriate for our global and industry needs and trends.

Results of the hydraulic model help the utility prioritize maintenance activities, optimize operational response, and plan for infrastructure growth or replacement, resulting in a prioritized capital program management (CPM).

Capital Program Management

Capital program management refers to the business practices and related technology solutions that help utility engineers, program managers, and construction managers control and manage the vast amount of information that capital improvement projects create. By their nature, capital programs are complicated. They are expansive in their vision, scale, budget, and time horizon. Traditionally, much of the management and control of these programs has been performed with Excel spreadsheets, which are unsuitable for the volume of data involved. Managing a couple of projects is relatively easy, so many of the smaller utilities can get by with using spreadsheets. However, as the portfolio of the utility's capital projects increases, effective and consistent capital program management becomes a much bigger challenge. The complexity of managing the data and information surrounding large capital portfolios requires a robust CPM software solution.

Capital program management software centralizes key project information related to project scope, cost, budget, estimates to complete, contractor payments, bid documents, as-built drawings, and project schedule. GIS enables the collection, storage, and use of multiple data sets to enable efficient communication and optimal decision-making in conjunction with CPM software. Most modern CPM software consists of the following major components:

- Schedule management and control to track one project, selected projects, or full project portfolios
- Document management to control bid documents, contracts, specifications, and drawings
- Cost control requiring methodical entry of cost data and automated tracking of changes orders
- Process automation to streamline repetitive and time-consuming processes
- Integration with back office ERP systems for near real-time cost information
- Performance reporting and dashboards to allow full transparency and informed decision-making

Modern CPM software can manipulate and leverage project-related data from hundreds of projects and standardize how the project information is managed, controlled, reported, and distributed. The software embeds industry best practices for properly managing the business processes associated with capital construction projects and makes the pertinent project information available to all key stakeholders. The utility's full project portfolio can be rolled up or reported separately by each project and displayed in GIS for effective and efficient capital program management oversight.

IT Planning

To improve by more effectively dealing with the various business challenges, utilities need to define and articulate their enterprise information technology (IT) vision. IT vision and supporting technology investments need to be closely aligned with the utility's top business strategies and workflows. The alignment of business strategy with the IT strategy is necessary in the utility industry for a variety of reasons: the new normal, climate change, rates, regulations, retirements, and the pressing need to be more cost-effective. The alignment of IT strategy with the utility's business strategy is a critical enabler for overall utility business improvement.

The development of an IT strategic plan helps to enable achievement of the utility business strategies and serve as the foundation for every future technology initiative. The IT strategic plan should address the following business needs:

- Align technology with the stated mission, vision, goals, and objectives
- Define a strategic direction and plan for all technology projects
- Implement best practices for managing technology investments, taking advantage of existing technology investments

- Delineate the key strategies for organizational and staffing structures to support technology
- Ensure that people, processes, and technology are aligned across the business

A successful IT strategic plan (Figure 5-6) must match your defined business objectives and gain the buy-in of the entire organization including management, operations, maintenance, finance, customer service, laboratory, and engineering personnel. Achieving improved organizational efficiency within a diversified and complex organization requires meaningful integration of information, business practices, and workflows throughout the organization.

To ensure that the IT plan addresses critical business needs, the first step of the project approach should include a thorough understanding of the departmental business initiatives, as well as any competing citywide or enterprisewide IT initiatives. The second step in developing an IT strategic plan is an inventory and assessment of the current business applications and associated business processes and workflows. Based on the improvement opportunities identified during the assessment phase, the next step is developing applicable strategies for reconfiguring, enhancing, or potentially replacing existing information systems. Once the strategies are identified and prioritized, a series of defined projects with budget, schedule, and resource needs can be planned for the next 3-, 5-, or 10-year horizon. The resulting IT strategic plan will thus ensure that technology investments are applied in such a manner that makes business sense across the entire organization.

Performance Management

In this period of economic uncertainty, utilities must focus on increasing their overall business efficiency to maintain their existing level of customer service at a reasonable cost. Performance measurement is the discipline of measuring the right things to effect a desired organizational change to achieve the desired business goals.

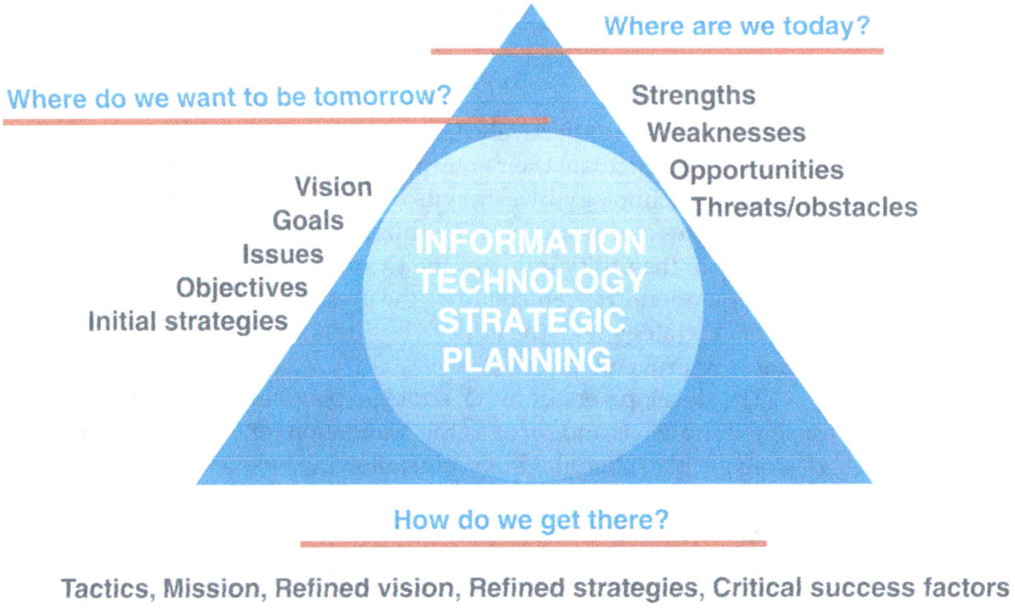

Figure 5-6 IT strategic planning

In order to be most effective, performance management programs in water utilities must be tailored to meet the agency's mission—ensuring a safe and reliable water supply, protecting the environment, maintaining public infrastructure, meeting customer service levels, and doing all at reasonable rates. Each agency's needs for improving its performance management program vary based on utility size, governance structure, range of services, and current organizational maturity. Accurate and timely performance data are at the heart of a successful utility performance management program. However, typically several organizational and technology challenges must be overcome to effectively utilize performance data, including:

- Key decisions are made without fully utilizing the objective-based data
- Key data from core business systems may be difficult to access and many times conflict
- There is an inability to objectively measure or trend progress toward the desired business goals
- There is lack of agreed-on performance-based target levels of service
- Ineffective performance trending tools inhibit communications with board or council members, regulators, ratepayers, and utility staff

There are many key performance indicators (KPIs) for a typical water utility. However, there is no single set of KPIs that applies to every water utility. Rather, agencies should begin with a smaller core suite of measures that support their stated mission, and then add measures as necessary to monitor their specific strategic initiatives and any pending management concerns. Each performance measure must be adequately defined, linked to a business strategy, monitored periodically, and effectively reported to all key stakeholders. Executive-level commitment is essential in order to sustain the utility's performance management program.

Water Audits

Nonrevenue water is a key metric used by water utilities to gauge the efficiency of their systems. An overall approach to understanding the sources of and solutions for nonrevenue water should be developed by utilities. While the inputs (water produced and purchased) and outputs (water billed) are generally known by the utility, less is known about the differences between these inputs and outputs.

The water audit methodology developed by the AWWA Water Loss Control Committee provides a valuable tool to understand the causes of water loss and provide a prioritized approach to addressing them. As more data are gathered about the various water uses and causes of lost water, a detailed picture of the utility's system begins to emerge (see Figure 5-7).

To aid in this effort, AWWA has developed a Microsoft Excel–based software program that manages the process and is available for free through the www.awwa.org site: AWWA Free Water Audit Software (sample screenshot shown in Figure 5-8).

For more in-depth information on this topic, please reference AWWA manual M36, *Water Audits and Loss Control Programs*.

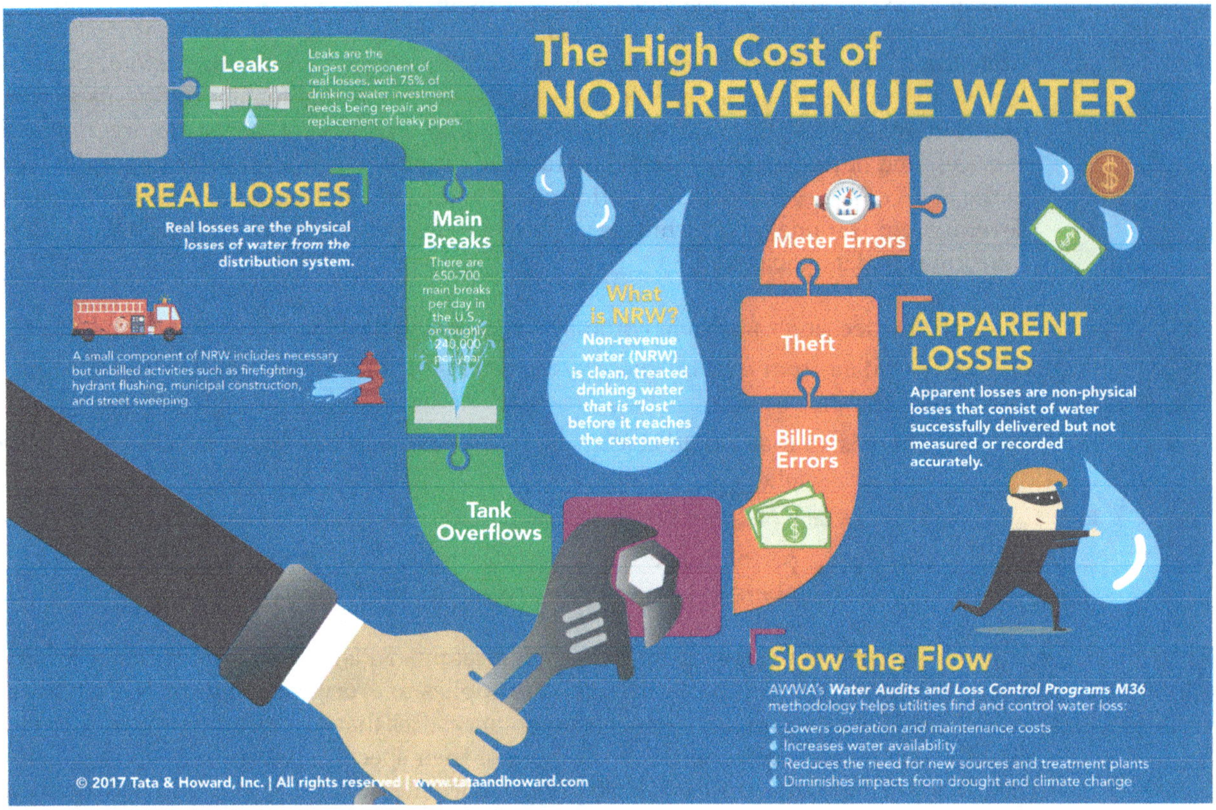

Figure 5-7 Sources of nonrevenue water

Source: Tata and Howard Inc. © 2017.

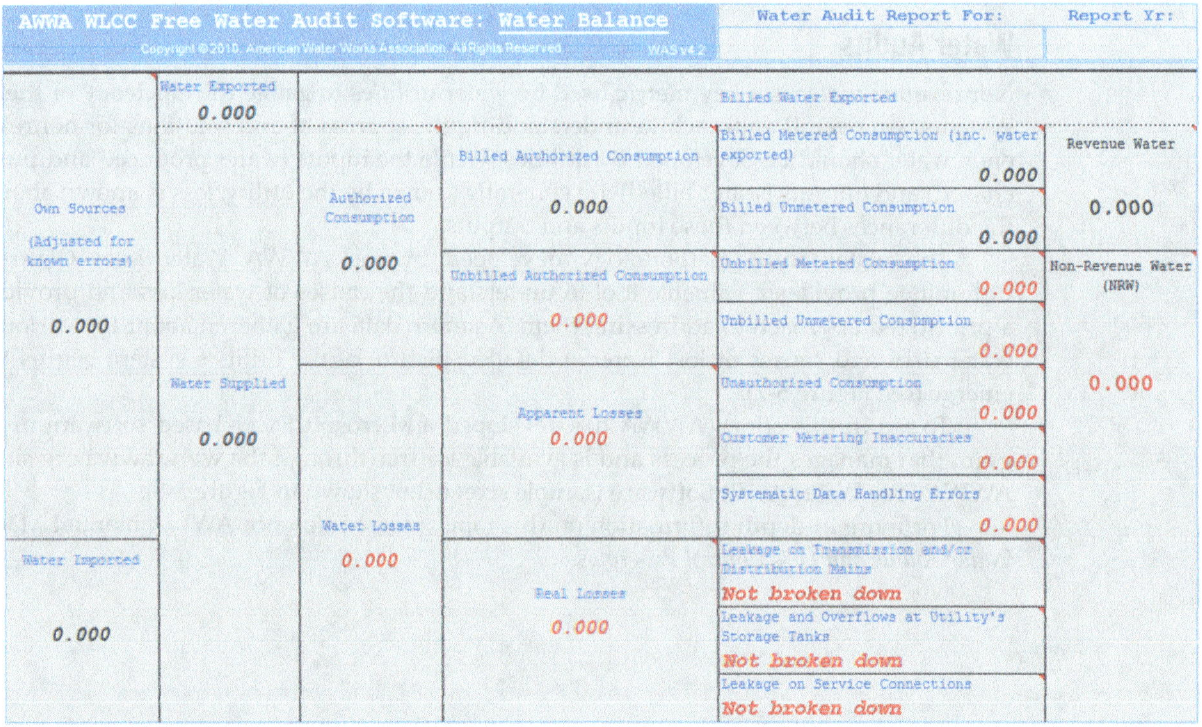

Figure 5-8 Screenshot from AWWA Free Water Audit Software

Source: AWWA Water Loss Control Committee 2010.

TRENDS AND THEIR IMPACTS

Big Data

The advancements in communications, data storage, and computing have led to an explosion of data in the water utility and in our daily lives. In 2007, just to put this into perspective, if we were to visualize the world's total data storage capacity in terms of standard compact discs (typically holding 730 MB of data), the stack of compact discs would have reached "from the earth to the moon and a quarter of this distance beyond" (Hilbert and Lopez 2011).

Here is an example of how this trend has played out from the water utility perspective. The Contra Costa Water District (CCWD) in Northern California provides treated and untreated water to approximately 500,000 people. CCWD utilitywide data storage capacity has grown an average of 33 percent per year since 2000, from 52 GB to over 6,000 GB (Figure 5-9; Morton personal communication) in 2016, while the population served has remained relatively stable. In addition, in 2010, CCWD implemented GIS functionality that resulted in a 400 percent increase in its data requirements. Interestingly, since the volume of GIS data peaked in 2012, CCWD has taken proactive steps to reduce its data footprint using data compression and implementing cloud storage.

Another example of this data deluge is seen when water utilities implement advanced metering infrastructure (AMI) programs. Consider a hypothetical utility with 60,000 metered water services. In years past, this utility may have read these meters once every quarter and saved these records for three years, resulting in 720,000 reading records. The move to an AMI system enables hourly (or more frequent) reading of these meters, resulting in nearly 1.6 billion records over that same three-year period (see Figure 5-10).

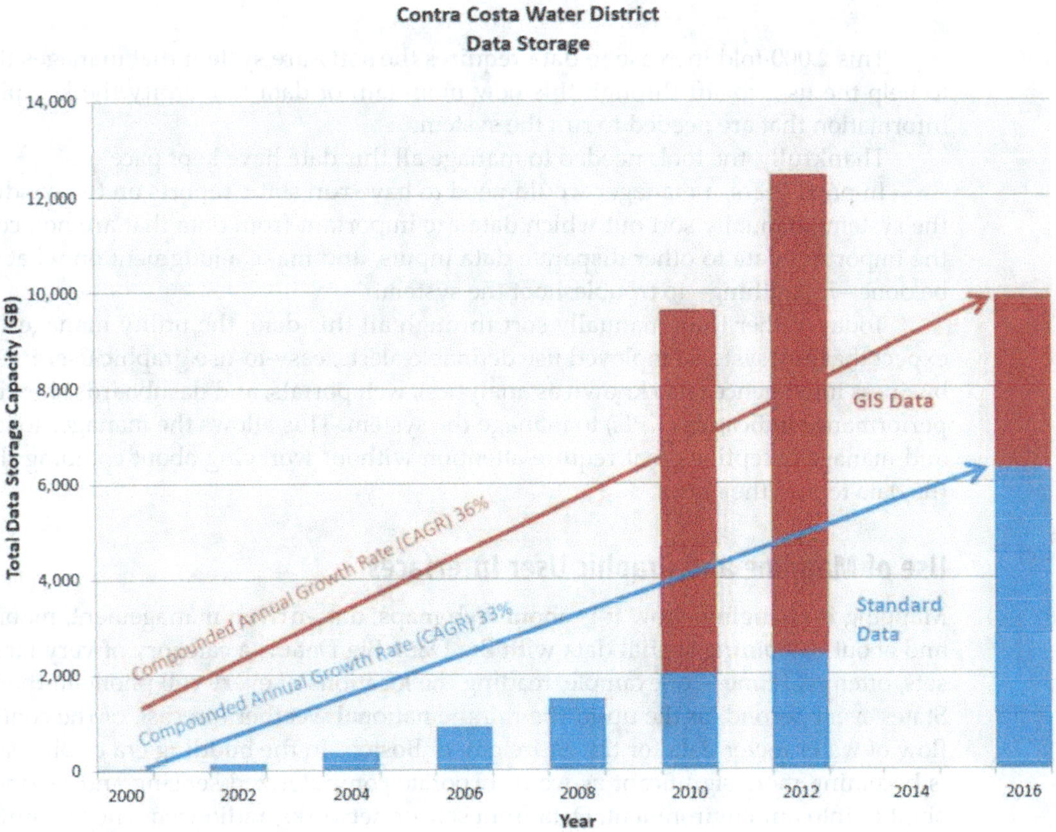

Figure 5-9 Contra Costa Water District (CCWD) data storage

Source: Contra Costa Water District. (Stephen J. Welch)

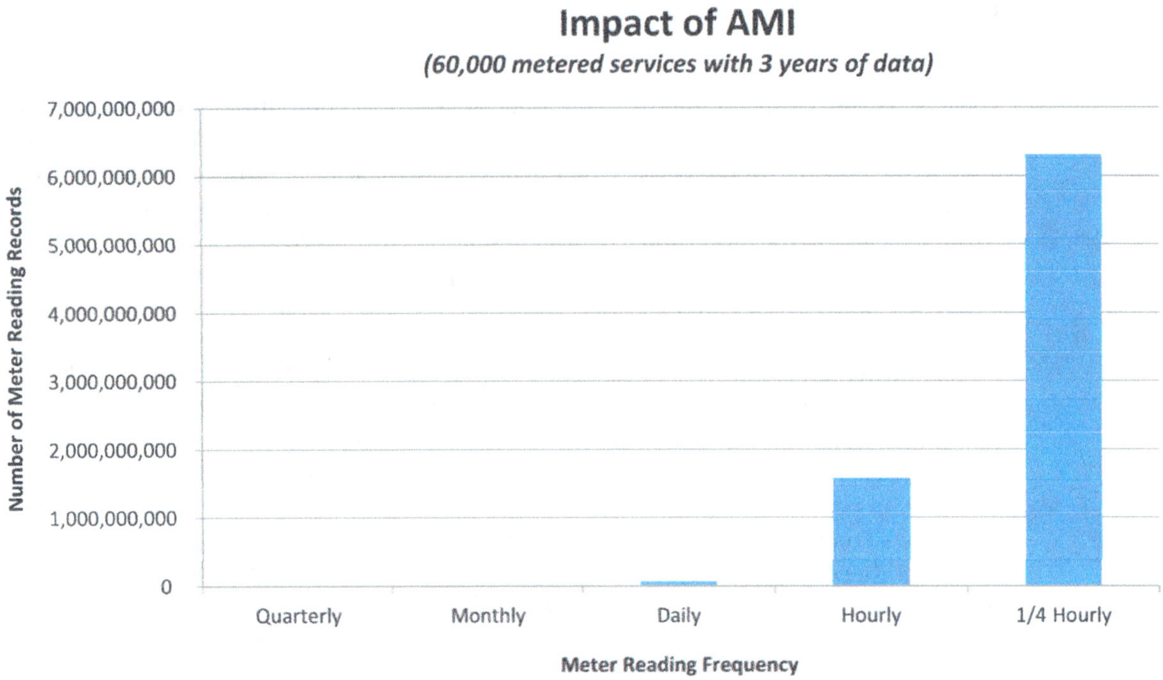

Figure 5-10 Hypothetical impact of AMI

Source: Neptune Technology Group Inc. 2013.

This 2,000-fold increase in data requires the software system that manages the data to help the user to sift through this new mountain of data to identify the key pieces of information that are needed to run the system.

Thankfully, the tools needed to manage all this data have kept pace.

In prior years, a manager would need to have run static reports on the condition of the system, manually sort out which data are important from data that are not, compare the important data to other disparate data inputs, and make a judgment on what should be done—if anything—to troubleshoot the system.

Today, rather than manually sort through all this data, the utility manager should expect that any system employed use definable alerts, easy-to-use graphic user interfaces, business intelligence (also known as analytics), web portals, and dashboards showing key performance indicators (KPIs) to manage the system. This allows the manager to identify and manage exceptions that require attention without worrying about combing through the data to find the issues.

Use of Mapping and Graphic User Interfaces

Mapping is changing—now it is about web maps, data-driven management, mobile GIS, and about combining spatial data with Big Data. Big Data is a category of very large data sets, often real time—for example, reading the locations of every cell phone in the United States every second, or the up-to-the-minute national weather forecast, or the continuous flow of water meter data for the entire city of Boston. In the budding era of Big Data, GIS is becoming more significant as we incorporate computerized sensing and broadcasting abilities into our environment. Data from sensor networks, radio-frequency identification (RFID) tags, surveillance cameras, and geotagged social-media posts all have geographical components and in real time. We are witnessing an exponential growth in both the number of data capture methods and, perhaps more significantly, in the amount of data being generated and captured. Part of data-driven management is pulling together different forms of information that can yield new insights. GIS uses location as the common thread needed to free data from previously narrow contexts and make it available for everyone.

Mobile Computing

With the advancements of the personal computer (PC) in the 1980s followed by the rise of laptops in the 1990s, mobile computing has changed the way we interact with information and people alike. The last decade has seen the emergence of a new form factor: tablets and smartphones. As of the first quarter of 2013, the tablet PC had surpassed the desktop and laptop to become the most popular device for modern computing (see Figure 5-11).

As consumers, we typically look to new technology to improve our daily lives. The adoption of mobile computing within the utility is now beginning to mirror what is taking place in the consumer market. Utility workers are seeing the value of leveraging the devices they use every day to be more effective, engaged, and connected at work.

This represents a new era in utility data management. The transition from costly workstations and rugged laptops to lightweight, inexpensive tablet PCs and smartphones allows today's utility to realize the benefits of information and communication at levels not previously seen.

The Rise of Cloud Computing

Cloud computing is entering a mature stage in its evolution. All predictions show no abatement in the increasing adoption rate for cloud computing and tools that can be used in a software-as-a-service (SaaS) model. Cloud computing and SaaS can help level the playing field for access to increasingly complex technology and business solutions.

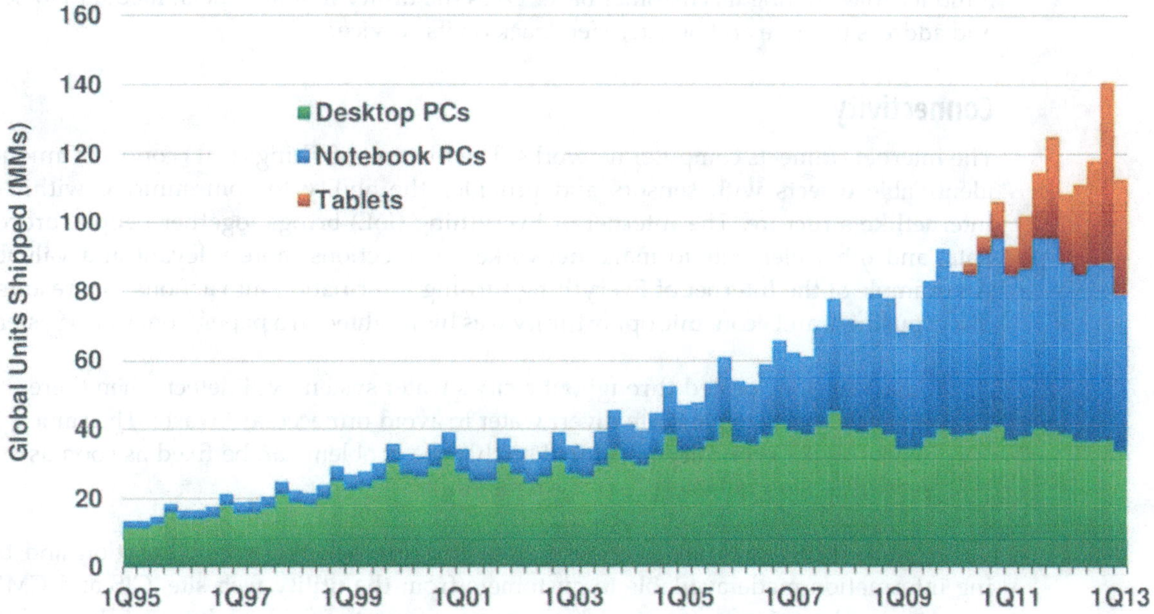

Figure 5-11 Global PC (desktop/notebook) and tablet shipments by quarter
Source: Huberty and Gelblum 2013.

Several factors make cloud computing an attractive approach for utilities of all sizes. The lower capital of systems based on a cloud-computing platform makes them extremely scalable. Ramp-up times for system deployment and support costs are also greatly reduced compared with typical host software approaches. Further, the utility no longer needs to manage version control as the software is typically updated automatically by the provider.

While system security will always be an issue as our world becomes increasingly connected, cloud-computing platforms are staffed with teams of people whose sole job is to keep the data protected and secure. Most water utilities lack the resources to come close to the level of system security that can be provided by experts who specialize in this field.

While still in their early phases of use in the water utility segment, cloud-computing platforms and software-as-a-service approaches will continue to gain ground, just as they have in the private sector.

Social Media

Social media have had and will continue to have a significant effect on how people consume data and make their voices heard. Increasingly we gather information through social portals and it was estimated in 2012 that people's social media use made up "20% of their time spent on PCs and 30% of their mobile time" (Nielsen Company 2012; the survey was "based on a representative sample of 1,998 adult (18+) social media users").

Equally important is the fact that social media are giving people a voice. This trend, often referred to as *democratization,* means that (rightly or wrongly) individuals have as much opportunity to influence public opinion as does the utility. Furthermore, because the individual is often seen as being unbiased, the comments of a few people on a social media site may have more influence over public opinion because the utility is considered to be pushing an agenda. In business, this trend has resulted in a shift in power from sellers to buyers, who now have a much stronger say than ever before in what they want.

Water utilities need to understand and participate in these trends and develop digital media strategies and policies to address them. Has the utility leveraged social media to build its voice among its customer base? Does the utility monitor social media to look for and address positive and negative feedback on its service?

Connectivity

The Internet connects computer networks. The Internet of Things (IoT) connects uniquely identifiable objects with sensors and provides the ability to communicate within an Internetlike structure. The Internet of Everything (IoE) brings together people, process, data, and other elements to make networked connections more relevant and valuable. An example of the Internet of Everything turning information into actions that results in direct business and economic opportunity was highlighted in a paper from Cisco Systems:

> Smart sensors located throughout a city's water system will detect when there is a leak and automatically divert water to avoid unnecessary waste. The same sensor will alert utility personnel so that the problem can be fixed as soon as resources are available. (Cisco Systems 2013)

Today, utility data are accessible from the Internet. Water consumption and billing information made available to customers from the utility web site, GIS and CMMS resources and services made available for access from the field via Internet Protocol (IP), and SCADA and AMI data available via web services (an emerging trend) are examples of utilities embracing the Internet to improve efficiency, reduce cost, and eliminate waste.

Automated metering infrastructure (AMI) that interconnects meters equipped with communication-capable sensors for gathering consumption and leak data in near real time is an example of the Internet of Things. Field crews alerted about a main-line break by a smart sensor in the network, then communicating with the office staff with their mobile devices (smartphones, tablets) to share field measurements, pictures, asset and event locations, and videos; an intelligent predictive model that recommends multiple what-if scenarios for closing the broken line with least amount of water loss and customer impact—all are the promises (and a near-term reality) of the Internet of Everything.

Interoperability

The *World English Dictionary* (Dictionary.com 2013) defines *interoperability* as "of or relating to the ability to share data between different computer systems, especially on different machines: interoperable network management systems."

Increasingly, the value that a utility can achieve from its systems is derived from its ability to share and analyze data across systems, often from different providers. But what is the best approach to enabling interoperability?

Figure 5-12 shows the basic framework used by many utilities to generate, transmit, capture, and analyze data.

Interfaces from most sensors are now standard and well documented. As long as the providers of the sensor and transmitter have an agreement that ensures forward and backward compatibility, the utility should be able to mix and match these devices.

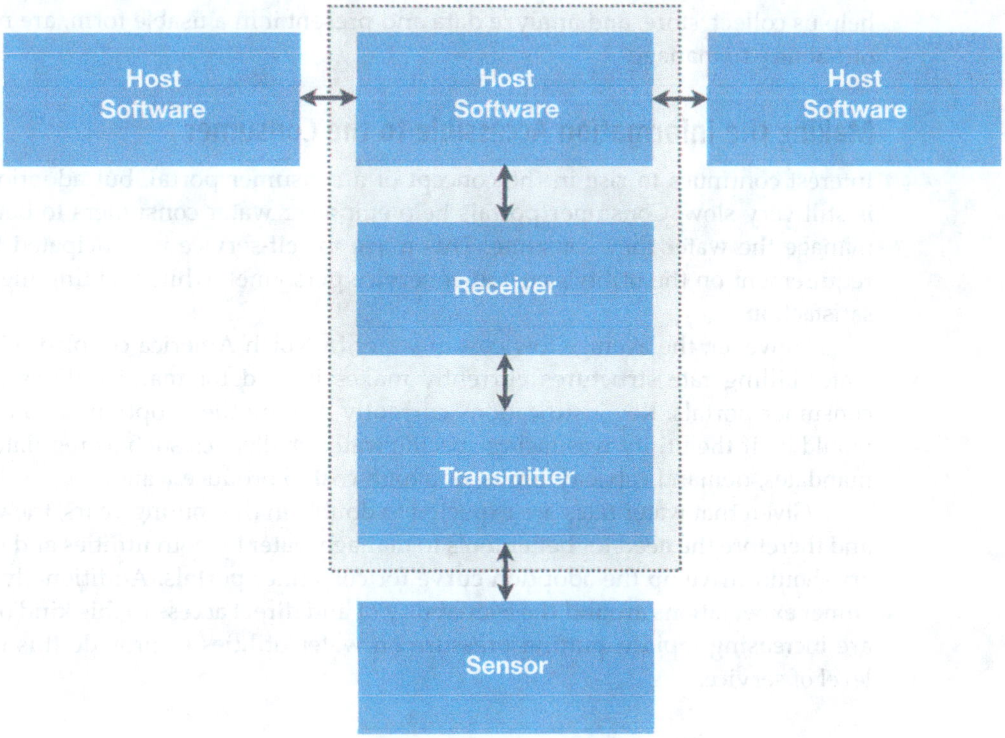

Figure 5-12 Architecture of sharing

It is common practice in the water utility sector that the transmitter, receiver, and host software be sourced from a single provider. The primary reasons for this convention are as follows:

- Communication methods and protocols vary greatly and "breaking" the system between any one of these components introduces the opportunity for lost or misinterpreted data
- Utilities generally prefer "one throat to choke" when it comes to critical systems; choosing a homogenous system eliminates the possibility for finger-pointing among different component providers
- Opening communication protocols among devices from multiple providers introduces a risk of cyber-hacking

An area where the water utility would benefit from improved interoperability would be the implementation of data communication standards among different host software systems. Sharing data at this level of the system architecture provides the most value with the fewest problems.

Growth of Business Intelligence

Business intelligence (BI), which may also be known as *analytics* or *business analytics,* is "an umbrella term that refers to a variety of software applications used to analyze an organization's raw data. BI as a discipline is made up of several related activities, including data mining, online analytical processing, querying and reporting" (CIO Magazine web site). The mountains of data that are now available require a new set of tools. No longer is it feasible to print an exception report and scan for anomalies. Business intelligence tools that help us collect, store, and analyze data and present it in a usable form, are now critical to our ability to manage.

Making the Information Accessible to the Consumer

Interest continues to rise in the concept of a consumer portal, but adoption of the idea is still very slow. Consumer portals help empower water consumers to better track and manage the water they consume. This move to self-service is anticipated to reduce the requirement on the utility's customer service personnel, while yielding higher customer satisfaction.

However, the average low cost of water in North America coupled with traditional water billing rate structures currently makes it hard for many utilities to cost-justify consumer portals. Key justifications currently driving the adoption of consumer portals would be if the utility was facing specific water challenges such as regulatory reduction mandates, demand/capacity limits, or a high cost to produce water.

Given that water rates are expected to double in the coming years, the value of water and therefore the need for better tools to manage water by both utilities and their consumers should drive up the adoption curve for consumer portals. Additionally, general consumer expectations around the availability of and direct access to this kind of information are increasing rapidly, putting pressure on water utilities to provide this new expected level of service.

Data Security

An interesting data security challenge faces many municipally run water utilities. Oftentimes, the utility IT function is integrated with or even completely dependent on the IT organization and infrastructure for the municipality. That dependence can mean that everything (utility data, city interdepartmental functions, billing, and more) is linked together. This is a critical point of failure and the utility should always strive be separated or have firewalls to "protect itself" from its own organization. This protection reduces the risk that changes or weaknesses in one part of the city's system adversely affect the function of the utility.

Trends

With current meter-reading technology moving steadily toward AMI, metering and data collection devices are becoming more sophisticated and therefore more interesting to cyber criminals (see Figure 5-13). This, in addition to traditional meter-reading systems becoming control systems, has led to an increased focus on ensuring that data and commands are being protected sufficiently and cannot be disrupted or modified for malicious uses.

The increased use of computing assets in the cloud and SaaS has pulled software and data that traditionally lived behind a utility's firewall to the Internet. The use of customer portals is giving the average utility customer a way to view usage data that goes far beyond a simple list of readings and can in many cases provide a profile of the way a customer uses water. This has added a number of additional "attack vectors" for cyber criminals to exploit as well as a host of new reasons to want to access the data. As a result, a utility needs to be proactive in making sure that its chosen system has the proper controls and structure necessary to protect both its system and the data that are collected.

As cybercrime has become more prevalent, sophisticated, and organized, the US government is taking an increasingly hands-on approach to ensuring that critical infrastructures are protected. Executive Order EO13636 is an early step toward building what will likely become a comprehensive framework used to set standards, methodologies, and processes that utility companies will need to conform to in the coming years.

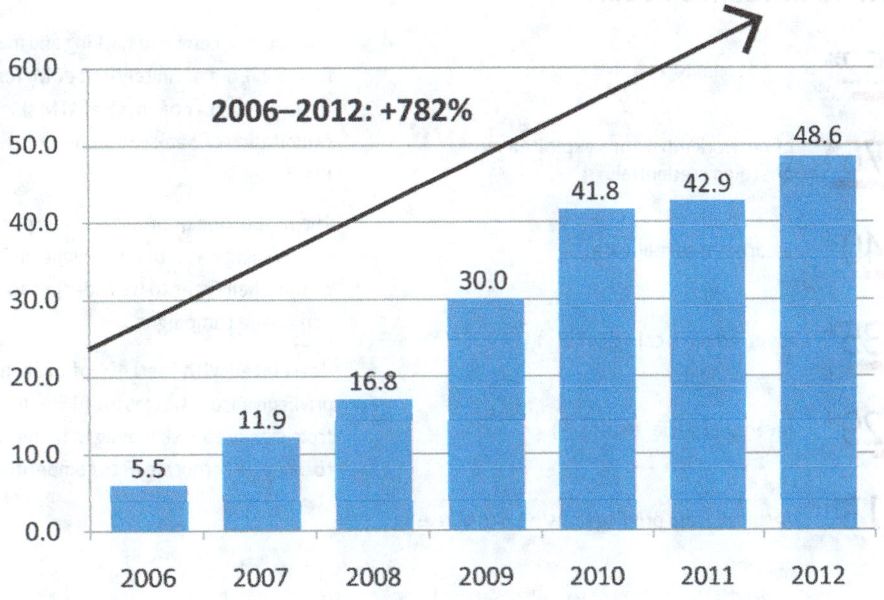

Figure 5-13 Cyberattack incidents reported by federal agencies

Source: US Government Accountability Office, US-CERT data.

Current Best Practices

Long gone are the days of a system using obscurity as a form of security. Modern systems need to make proper use of as many security measures as possible. From the beginning, systems need to be designed and implemented with security in mind and not have security added after the fact. All points of communication or access in a system need to be evaluated when determining overall security level of the system.

Utilities should ensure that their chosen vendors have implemented proper encryption for data, both in motion and while sitting still. All users and devices should make use of authentication routines before being allowed to interact with any part of the system. The use of firewalls can help keep data safe, but an intrusion detection system can be a more effective tool to identify and stop malicious activity.

Do not underestimate the value of installing a good virus scanner and keeping it up to date with its latest updates. Many times, the initial foothold an attacker gains is through a simple malware program, and though not perfect, virus scanners can help reduce the risk of this happening. Coupling a good virus scanner with a program to ensure that the operating systems of all computers are updated with patches provides a good start to ensuring that internal computers are not working against the rest of the protections built into the system.

Even with the proper use of encryption, authentication, and other security techniques, the weakest link of any system still needs to be addressed. According to Verizon annual report that summarizes the makeup of reported data breaches across all industries (Verizon 2013), the use of social engineering attacks increased four times over what was seen in the prior year (see Figure 5-14). It is of critical importance that all employees of the utility are trained to recognize social engineering attacks and take the necessary steps to keep attackers out. A simple click on a malicious link in an email or unknown document can offer all of the access to your network that a hacker needs to steal data or take control of a system without raising awareness. Studies have shown that if attackers can

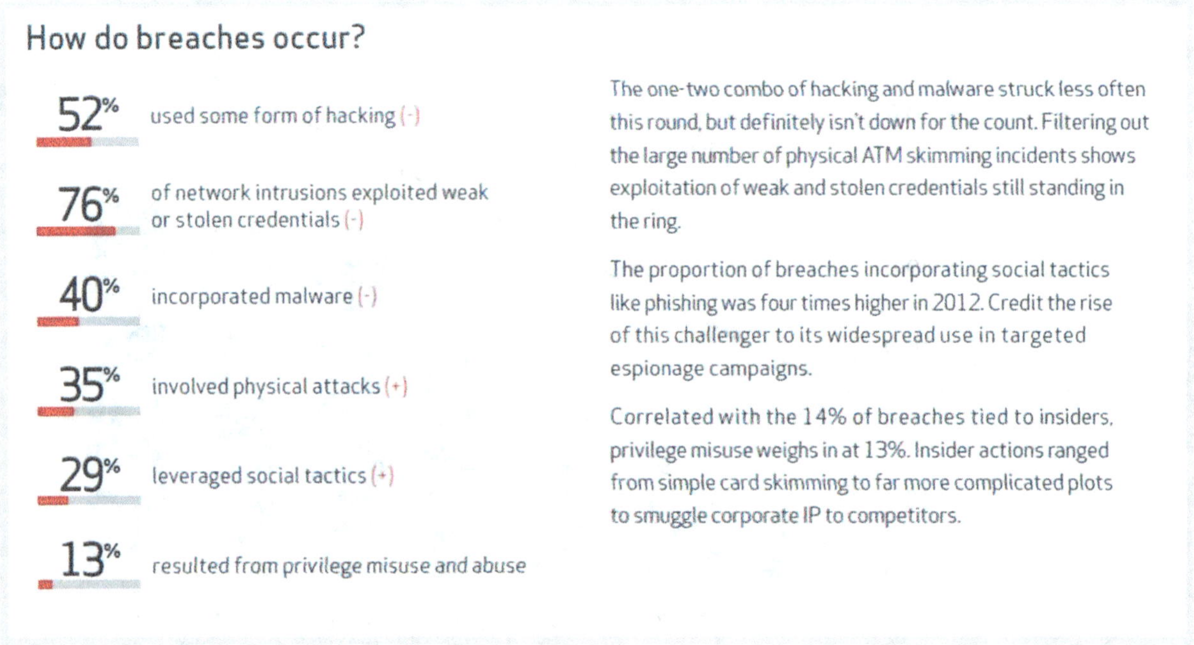

Figure 5-14 Sources of data breaches

Source: Verizon 2013.

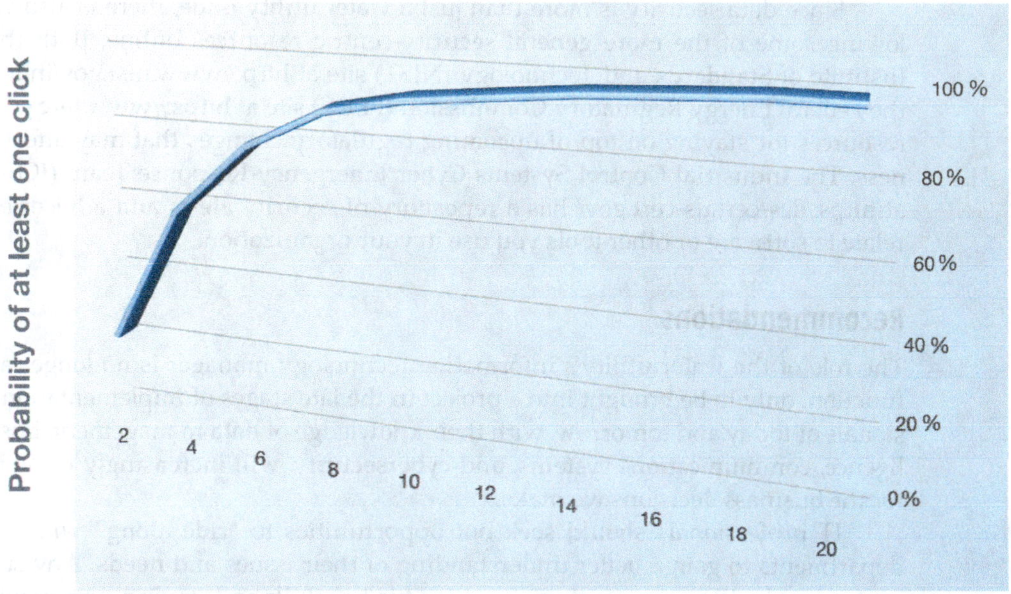

Figure 5-15 The inevitability of the click

Source: Verizon 2013.

send emails to between 14 and 20 different employees of an organization, they are nearly guaranteed that someone will click on a malicious link or open an infected attachment (see Figure 5-15).

Users should also be encouraged to use pass phrases instead of traditional passwords. A good pass phrase should consist of multiple words that are not normally used together in normal conversation, music, or literature and do not contain personal information. For example, using the words *correct, battery, horse,* and *staple* along with some punctuation could create a password of "CorrectB@tteryHorseStaple!"; such a password would be much more secure than a traditional-style password of six to eight characters, while being easier to remember.

Once a system has been installed, it is a good idea to bring in a security firm that is capable of executing a penetration test against the system to simulate an attack by hackers. Although these types of tests are expensive, they can help to identify any possible holes in the implementation and give a real-world idea of how well a system will withstand an attack. Most penetration testers will provide suggestions on how to remediate any issues that are found during the testing, and many will retest the issue after it has been resolved to help ensure that the fix was properly applied. Working with relevant vendors and a penetration tester will ensure that there are no unknown back doors into the system.

Tools That Utilities Should Use to Stay Informed

Though somewhat focused on the SCADA side of the utility, the WaterISAC security network (https://portal.waterisac.org) provides an excellent set of resources that can be used to keep up with current events in the water sector as well as notable security holes found in systems that may be in use at your utility. AWWA also has a Security Preparedness Group that acts as a liaison between the water utility industry, the federal government, and other security agencies to provide guidance to water utilities on best practices for cyber security. Reference material can be found at http://www.awwa.org/legislation-regulation/issues/utility-security.aspx.

Since data security is more than just a water utility issue, there is also value in following some of the more general security-centric resources online. Both the National Institute of Standards and Technology (NIST) site at http://www.nist.gov/index.html and the Federal Energy Regulatory Commission (FERC) site at https://www.ferc.gov/ are great resources for staying on top of upcoming regulatory changes that may affect your business. The Industrial Control Systems Cyber Emergency Response Team (ICS-CERT) site at https://ics-cert.us-cert.gov/ has a repository of security alerts and advisories that may relate to software or other tools you use in your organization.

Recommendations

The role of the water utility's information technology manager is no longer a backroom function, only to be brought into a project in the late stages of implementation. IT professionals of today and tomorrow, with their knowledge of data management, business intelligence, communications systems, and cybersecurity, will increasingly provide the basis for the business decisions we make.

IT professionals should seek out opportunities to "ride along" with other utility departments to gain a better understanding of their issues and needs. Any successful IT professional will be constantly sharpening his/her skills and staying current with technology changes and social trends.

Managers in other water utility departments should involve IT early and often in their consideration and implementation of new systems.

The water utility can no longer be considered a group of silos, each with its own systems and information. Utility general managers should look at the utility from a holistic perspective to understand how information that is generated in one part of the utility could be shared among other departments to add value for the entire organization.

REFERENCES

AWWA. 2012. M32. *Computer Modeling of Water Distribution Systems,* 3rd ed. Denver, CO: AWWA.

AWWA Water Loss Control Committee. 2010. AWWA Free Water Audit Software. 2010. Denver, CO: AWWA.

CIO Magazine web site, http://www.cio.com/article/40296/Business_Intelligence_Definition_and_Solutions.

Cisco Systems. 2013. http://www.cisco.com/web/about/ac79/docs/innov/IoE.pdf. Accessed on Oct. 1, 2013.

Deloitte Touche Energy and Resources. 2012. *Water Tight 2012: The Top Issues in the Global Water Sector.* New York: Deloitte.

Dictionary.com. 2013. *World English Dictionary,* Collins, http://dictionary.reference.com/browse/interoperability?s=t. Accessed on Sept. 20, 2013.

Federal Energy Regulatory Commission (FERC) web site, https://www.ferc.gov/.

Hilbert, M., and P. Lopez. 2011. The World's Technological Capacity to Store, Communicate, and Compute Information. Published Online Feb. 10, 2011, *Science* 1 April 2011. 332(6025): 60–65, DOI: 10.1126/science.1200970.

Huberty, K., and Gelblum, E., Morgan Stanley Resesearch, Gartner. 2013. "Internet Trends D11 Conference." Published by Kleiner Perkins Caufield Byers; www.kpcb.com/file/kpcb-internet-trends-2013. Accessed Aug. 16, 2013.

Industrial Control Systems Cyber Emergency Response Team (ICS-CERT) Web site, https://ics-cert.us-cert.gov/.

Jernigan, W., and Cavanaugh, S. 2013. "AWWA Free Water Audit Software © (v 4.2) Demonstration." http://www.ncsu.edu/wrri/pdfs/pastevents/awwa102011/WillJernigan.pdf. Accessed Aug. 9, 2013.

Morton, J., Manager of Information Systems, Contra Costa Water District. Personal communication.

National Institute of Standards and Technology (NIST) web site, http://www.nist.gov/index.html.

Nielsen Company. 2012. "State of the Media: The Social Media Report, 2012." p. 4.

Verizon. 2013. "The 2013 Data Breach Investigation Report." http://www.verizonenterprise.com/DBIR/2013/. Accessed on Sept. 20, 2013.

WaterISAC web site, https://portal.waterisac.org.

CONTRIBUTORS/AUTHORS

- Lori Armstrong—Global Industry Manager Water/Wastewater, Water Resources, Esri, larmstrong@esri.com
- Dave Hanes—Director of Strategic Marketing, Neptune Technology Group, dhanes@neptunetg.com
- Paul Hauffen—CEO, IDModeling, paulhauffen@idmodeling.com
- Kevin Koshko—Enterprise Solutions Director, IDModeling, KevinKoshko@idmodeling.com
- John Sala—Director of Systems and Software Marketing, Neptune Technology Group, jsala@neptunetg.com
- Doug Spiers—Vice President of Western Region, Westin Engineering, doug.spiers@we-inc.com
- Sreeresh Sreedhar—CTO, IDModeling, sree@idmodeling.com
- Darren Sulfridge—Principal Security and Software Architect, Neptune Technology Group, dsulfridge@neptunetg.com

This page intentionally blank.

Chapter 6

Taking Care of Your Assets: A Commitment to Infrastructure Management

This chapter contains a description of the essential ingredients of the water–wastewater industry approach to creating strategic asset information and recommendations, a critical step to informing the future planning of utility management.

Water and wastewater utilities are asset-intensive organizations. They rely on a complex network of pumps, pipes, storage tanks, treatment systems, and related infrastructure assets to convey water from source to customer. Ensuring that these assets operate in a manner that produces the desired level of performance at the lowest life-cycle cost is a big part of utility management. Achieving this objective is often a significant challenge. Utilities never have enough resources to maintain all their assets in excellent condition and prevent every failure. They must constantly balance asset-related risks, budgets, and service levels to maximize the utility's resources, while minimizing costs to customers. If done well, a good balance creates optimum value for customers.

The importance of infrastructure assets to a water or a wastewater utility has been recognized since the beginnings of public water supply, although the practices associated with managing assets have evolved considerably. In the early days of public water supply prior to 1900, the focus was on installing assets to address fire protection and immediate health concerns like cholera and typhoid outbreaks, and providing a rapidly growing population with basic services. Little attention was given to topics like maintenance planning, asset sustainability, and risk management. Service expectations were understandably low, since many customers had little or no history with access to piped water.

Customers demand far more of their utilities today than in the past. Meeting these demands requires higher levels of asset performance. The new paradigm for managing assets effectively requires several elements. These include

- **Inventory**—Understanding what assets the utility is responsible for and their location and current condition. Defining criticality and performance requirements is essential.
- **Levels of service (LOS)**—Utility managers must know what LOS are required by regulations, what levels customers desire, and the related costs of these different service levels. Getting value from assets is about making sure they perform as desired at the right cost.
- **Information management**—Systems, practices, and processes are needed to capture data about assets and to process the data into information that can be used to make decisions about maintenance, rehabilitation, and replacement. Geographical information systems (GIS) and computerized maintenance management system (CMMS) are among the array of tools for storing and managing asset information.
- **Risk assessment**—Utilities should determine where they are susceptible to risk, how great the risk is, and methods to mitigate that risk.
- **Capital planning and budgeting**—Rehabilitating and replacing assets are a major part of an asset's life cycle, so these activities must be planned for, managed, and budgeted accordingly.
- **Operations and maintenance (O&M) practices**—How an asset is operated and maintained will have a significant impact on an asset's performance, longevity, and life-cycle costs.

Each of these elements fit together integrally. They will be explored in greater depth in this chapter.

EFFECTIVE UTILITY MANAGEMENT

Managing assets efficiently and cost-effectively is a critical component of water utility management. Effective utility management (EUM), a framework that is embraced by the leading water and wastewater industry organizations, considers *infrastructure stability* as one of its ten attributes of effectively managed water sector utilities. Several of the other ten attributes, such as *operational resiliency* and *operational optimization*, rely heavily on sound, efficiently functioning assets. EUM offers guidance, recommendations, and support information on managing assets.

FORMAL ASSET MANAGEMENT

Many elements are required to manage infrastructure assets effectively. Many of the practices and approaches are captured in the body of knowledge collectively referred to as *asset management*. Asset management represents a formal approach to managing assets, but leverages many well-vetted and common-sense practices that are recognizable to all water utility managers. The concept of formal asset management originated in Australia and New Zealand (see New Zealand Asset Management Support, www.nams.org.nz) but spread to Europe and the United States over the past two decades. Asset management sets forth a decision-making framework that is driven by asset information such as performance data, maintenance history, asset condition, and consequence of failure to make decisions based on risk, which for asset management purposes is a combination of the probability of an asset failing or not meeting its performance requirements and the consequences of that failure to the utility and its customers.

Many sources describe in detail the extensive formal elements of asset management practice. The *International Infrastructure Management Manual (IIMM)* is one of the best global sources of information. The *IIMM* document is now in its 5th edition and covers a wide array of topics from developing an asset inventory to capital planning. While the *IIMM* is an invaluable reference, the guidance provided is influenced by the operating conditions of many foreign utilities. Larger public and quasi-public utilities are common in many of the founding asset management countries. These foreign utilities often operate under a more uniform national regulatory framework that encompasses environmental, financial, and service standards collectively. These conditions are different from what many US utilities experience, so parts of the *IIMM* guidance must be adapted for US utilities. In the United States, water quality regulations are largely defined at the national level through the Safe Drinking Water Act (SDWA) under the jurisdiction of the USEPA, and service and financial requirements are defined at the state, county, and/or local municipal levels. Financial and service requirements can vary dramatically from state to state. Adapting asset management practices to fit a specific utility is imperative.

For utilities to make the most of their investment in each asset requires anticipation of risk, anticipation of the need for replacement, and an overall reduction in the cost of capital, even at the micro level. AWWA's modeling software, Buried No Longer pipe need predictor, is designed to help utilities know where to replace, when to replace, and how much replacement or expansion is required in the system. Cost estimations are based on a mix of pipe types, pipe age, system size, and location.

SIMPLE, the Sustainable Infrastructure Management Program Learning Environment, sponsored by the Water Environment Research Foundation (WERF) and Water Research Foundation (WaterRF), provides asset management information that is more customized for US utilities. SIMPLE is endorsed by the USEPA. *Implementing Asset Management: A Practical Guide* by the Association of Metropolitan Water Agencies (AMWA), the National Association of Clean Water Agencies (NACWA), and Water Environment Federation (WEF) is another excellent guide to asset management practices that is very useful to US utilities and is endorsed by the USEPA. For smaller water systems the USEPA has created the Check Up Program for Small Systems (CUPSS). This interactive software tool and accompanying guidance document provide a great source of information on asset management. These documents and a host of other guides, publications, references, and case studies put out by the USEPA, major industry sources such as AWWA and WEF, and water research organizations like WaterRF and WERF provide a plethora of information on asset management practices.

Whether formal asset management practices or more informal infrastructure management principles are adopted, all utilities must employ a deliberate and systematic approach to managing their assets. This approach must consider the concepts of inventory, LOS, information management, risk assessment, capital planning and budgeting, and operations and maintenance practices described previously.

Additionally, each water and wastewater utility beginning an asset management approach benefits from referencing the hundreds of voluntary standards of minimum requirements, materials, equipment and practices used in water and wastewater treatment and supply. AWWA has developed these standards, which are used by thousands of manufacturers, distributors, and facilities worldwide to ensure the highest-quality products and service.

Asset Inventory

Managing assets effectively requires using information to make good decisions. An asset inventory is one of the primary sources of this information. Developing an asset inventory (age, size, location, nameplate data, etc.), or an *asset registry* as it is sometimes called, is an important step in properly managing assets. The USEPA and other sources consider the inventory a first step in formal asset management (see Figure 6-1). It is difficult at best to manage assets if a utility does not know what assets it owns or is responsible for.

Most basic asset data will come from records, maps, databases, and reports. Often the data will be scattered around the utility in different locations. Some of it will be paper-based and other data will be contained in electronic databases such as a GIS or a customer billing system. There are likely to be gaps in information. Some gaps will be easy to fill, while others may require copious amount of time or money to address. The cost in resources to fill these information gaps must be weighed against the expected benefit from the data to determine what level of data gathering is justified.

Developing an inventory can help identify and quantify asset performance problems, funding and manpower constraints, and other utility strengths and weaknesses. The inventory can be linked to other information sources, typically through a unique numeric asset identifier, such as the volume of water treated and pumped, customer billing data, and work order information. These records provide data on the work activities' performance, level of customer satisfaction, number and type of complaints (by category), and other system information. The data enable managers to make informed decisions concerning the types and levels of service offered. While complex analysis tools help to discover a utility's problems, simply plotting records as an overlay on the utility system map or as-built drawings, or just viewing the data in aggregate, can tell a utility manager a great deal about what is happening. Knowing a pipe's age, material, leak record, and installation environment, for example, can help predict when it will need to be rehabilitated or replaced. Although many factors can cause a pipe to fail—such as installation error, soil conditions, water quality, or pressure surges in the distribution system—these characteristics can help narrow the investigation and save a utility manager valuable time.

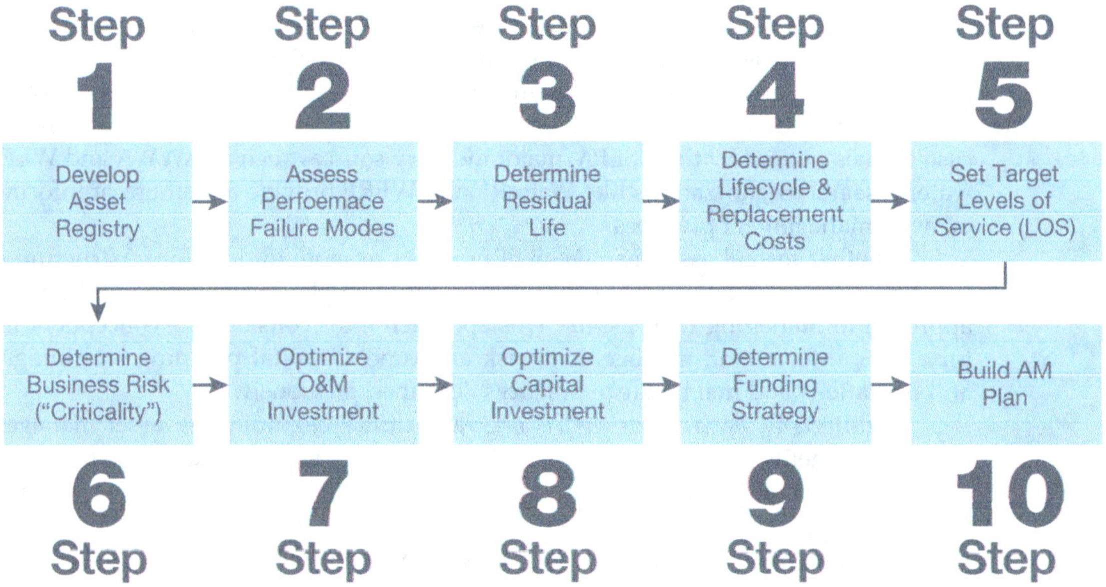

Figure 6-1 USEPA's 10-step asset management (AM) approach

Source: US Environmental Protection Agency

- City Water System
  - — Water Treatment Plant
    - • Pretreatment
    - • Sedimentation
    - • Filtration
      - ~ Filter 1
      - ~ Filter 2
      - ~ Filter 3
      - ~ Filter header piping
    - • Disinfection
  - — Water Transmission and Distribution System

- City Water System
  - — Water Treatment Plant
  - — Water Transmission and Distribution System
    - • Pressure Zone A
    - • Pressure Zone B
      - ~ High Street Neighborhood
      - ~ West Side
        - * Segment 1
        - * Segment 2
        - * Segment 3
    - • Pressure Zone C

- City Water System
  - — Aqueduct
    - • Watershed A
      - ~ Reservoirs
        - * R1
        - * R2
        - * R3
      - ~ Pump Stations
        - * PS1
        - * PS2
        - * PS3
    - • Watershed B

Figure 6-2 Sample hierarchies of assets at a water system

Organizing asset inventory information in a logical and systematic way will ensure people can readily gain access to it. Many utilities organize their data into an asset hierarchy where individual assets are grouped by process or location, for example, and then subsequently grouped into large collections representing a treatment plant or areas of a distribution system such as a pressure zone. Figure 6-2 provides a sample asset hierarchy for a water system. Grouping like assets allows utility personnel to quickly find individual assets in a database. In addition, it allows for the comparison of asset information across the organization.

Levels of Service (LOS)

Providing service that meets or exceeds customer expectations and is compliant with regulations is the primary goal of any utility. Utilities need specific, measurable service standards against which they measure themselves. These goals should be SMART—specific, measurable, achievable (or applicable), relevant (or realistic), and time based (specific date/deadline). Levels of service (LOS) should form the basis for determining appropriate practices, approaches, and funding for capital renewal and maintenance activities. Utility managers must know what levels of service are required by regulations, what LOS customers desire, and the related costs. Producing water that complies with SDWA and state regulations requires treatment equipment that operates properly—removing contaminants out of the source water—and transmission and distribution assets that protect the water while it is being conveyed to customers. This scenario is not possible without assets that perform as intended. On top of regulatory requirements, customers have quality and consistency expectations about water service. These expectations include reliable water pressure, a minimum number of pipe and service failures, and good-tasting water. In many cases, the performance level of assets required to meet customer quality and consistency demands is higher than to address water regulations.

Establishing LOS standards and goals starts with measuring what is currently being provided to customers. Utilities should understand how many breaks they experience per year, their levels of nonrevenue and unaccounted-for water, the accuracy rate of their bills, and many other LOS measures. Understanding what a utility is currently providing is a starting point for modifying service level goals. Good data and appropriate LOS will help to define how much investment is needed for pipe renewal, maintenance activities, and many other asset management aspects. If a utility does not know what customers want for service and how much they are willing to pay for different LOS, it is very difficult to effectively manage the water system. Many utilities assume they know what LOS the customers want and how much they are willing to pay to achieve them. Others have tried to get by without any kind of defined strategy. Knowing what needs to be achieved in terms of service allows utility managers to clearly define resources and actions to support goals.

Defining appropriate LOS is often a significant challenge for utilities, because it almost always means compromise. Customers typically demand exceptional service at little or no cost. Establishing LOS goals is done through an iterative process where the utility defines regulatory requirements, measures current LOS provided, and discusses with stakeholders desired service levels. Utilities need to communicate with customers and other stakeholders to determine what long-term LOS will achieve the best balance between cost (rates), risk, and levels of service desired by customers. An LOS discussion involves public outreach, which includes clarification of a utility's understanding of customer service desires. This can be done through discussions with appointed or elected utility boards, customer focus groups, and interviews with key stakeholders, or through more elaborate means like surveys, studies, and testing. Establishing LOS should be validated both with the utility's appointed or elected governing entity (board, council, commission, etc.) and with members of the public. Many utilities perform outreach that informs customers of the utility's service goals and their current performance, which is helpful for stakeholder (customer) acceptance and buy-in. Many find that developing a multimedia communication strategy (web site, Facebook, text messaging, emails, mail, newspaper ad, editorial, etc.) is more effective than any single communication method.

Many sources define regulatory requirements, including USEPA and state drinking water program web sites. The *Federal Register* is the definitive source. Measuring current LOS provided requires reviewing operational records, maintenance data, and anecdotal information from utility staff and stakeholders. The EUM framework offers guidance on the areas where many utilities try to measure their LOS. AWWA's utility benchmarking program and the International Water Association (IWA) *Water Utility Benchmarking* publication provide guidance on what LOS criteria utilities commonly track. Examples include number of main breaks per 100 miles of water mains, cost per million gallons produced, average time to restore water service, and so on. Table 6-1 presents examples of common levels of service and targets as first identified in 2005.

Once the basic inputs are established (inventory, reporting/records—repairs/outages/customer calls, performance metrics/LOS), asset management can evolve to include more analysis. The analysis can provide more specific repair plans, reports, and budget strategies. Some analysis software programs can create a work plan based on funding levels and show how funding certain projects will affect not only the system but also what customer functions are at risk. Information like this can provide the necessary justification to schedule work and increase funding. Nevertheless, even with the most brilliant analysis program, the insertion of technical expertise (human interaction) is needed because not all analysis produces options that are feasible for implementation.

Table 6-1 Examples of common levels of service and targets

Service Area	Performance Measures	Performance Targets	Calculation
Asset preservation, development, and consolidation	Preservation—the assets are being properly maintained	Not more than 10 breaks and leaks per 100 mi of distribution*	Planned maintenance ratio (hours) = (100)(hours of planned maintenance)/(hours of planned + hours of corrective maintenance)
Service quality (including aesthetics, reliability, responsiveness, and capacity)	Water outages are efficiently remedied	All break and leak repairs were completed within four hours of identification*	Unplanned disruption rate = (1,000)(number of water service disruptions)/(number of facilities connected)

*Performance target source is Lafferty and Lauer 2005.

Information Management

Utilities can capture vast amounts of data about their assets. For this information to be used effectively, it must be housed in systems that can help capture, manage, analyze, and distribute it. Without these systems, data cannot be used to make good decisions about maintenance, rehabilitation, and replacement of assets. There is a vast array of tools and systems for managing information. They range from simple paper files to complex cloud-based databases that are interconnected. The appropriate system depends on the scope of the utility and its data management goals.

Asset information can be scattered across a utility in many locations. Most utilities find that organizing information in four connected data repositories is efficient. These four data repositories and their descriptions are as follows:

- **Work information**—The information associated with maintaining assets (labor, equipment, materials, costs, etc.) is often contained on work orders. Work orders can be written documents or can be electronic records populating the database of a computerized maintenance management system (CMMS). A utility needs an organized and logical way, whether written or computerized, of storing information about what it requires in financial and other resources to maintain assets.
- **Asset information**—The size, age, location, nameplate information, and other defining attributes of assets must be stored so that the data can be easily accessed and used. In the past, most utilities stored this information in paper records such as plans, specifications, manuals, and so on. Many still do. A growing number of utilities now store information in a variety of databases from geographical information systems (GISs), which are highly valuable at displaying and analyzing data geospatially, to Microsoft Excel spreadsheets and elaborate document management systems. Regardless of the storage medium, there must be a logical and consistent system for locating, updating, and using the data.
- **Financial information**—The costs associated with building, maintaining, and operating assets are often found in financial records or in financial information systems (FISs). Effective utilities have methods of linking this data with their other data repositories so they can determine where they are spending their money and effectively plan and budget.
- **Asset monitoring and analysis**—A wide array of systems is on the market for monitoring assets and using asset information to make decisions. AWWA and other sources have manuals and standards documents devoted to the various monitoring and analysis systems. Supervisory control and data acquisition (SCADA) systems, for example, are common at utilities. They allow huge amounts of data on assets to be effectively viewed and managed through simplified graphical interfaces. Other monitoring systems include alarm packages, security systems, measurement devices, and so on. A wide array of analysis systems is also available. Some of the more common are hydraulic system models, meter reading systems, and GIS-based software.

Risk Assessment

The failure of an asset to perform as required can have a range of impacts on a utility. There are catastrophic failures ranging from the well-publicized water main breaks that appear on the evening news to the minor failures that can be readily fixed with little impact. Each of these events carries a risk to the utility, which can be characterized in terms of economic (financial), environmental, regulatory, and social (political) impacts. Risk in formal asset management terms is the probability of an event occurring and the cumulative consequences of that event. It drives decisions within an asset management framework. A utility must have an understanding of its most significant risks as well as its risk tolerance before entering into a discussion of managing assets. Identifying, measuring, and managing a utility's financial, legal, operational, and environmental risks related to investment in infrastructure pose a complex challenge, since multiple variables can occur and they must be considered with respect to many types of risks encountered.

A firm grasp on risk can help a utility take constructive action to minimize the impacts on customers of asset failures. It can also reduce the associated costs inherent in the business decision-making processes, lessening the likelihood that unexpected expenses will impact capital improvement or operational budgets. A utility needs to identify the parameters of risks (critical customers, sensitive infrastructure, economic zones, etc.) that are important to it. Once risks are evaluated and quantified, an accompanying evaluation criterion can be developed to help drive decisions.

Capital Planning and Budgeting

Effective capital planning and budgeting are needed to ensure the sustainability of assets in the short term and the long term. Both help with delivering the desired LOS at the lowest possible life-cycle costs while managing overall risk. Like most of the other aspects of managing assets, they require a careful balance among the competing elements of money (funding), service, and risk. Without the proper balance, negative consequences such as wasted resources, higher costs, and unnecessary risks occur. Effective capital planning has many components that cannot be adequately covered in this manual (please see AWWA M1 and AWWA M29 for more extensive coverage). Those aspects that apply to a utility's assets must be structured and should apply risk-based approaches derived from asset management practices.

The traditional capital improvement plan (CIP) model as shown in Figure 6-3 is still practiced by many public organizations. The process typically involves utility representatives developing a wish list of projects based on their experience and knowledge of their group's assets, which are then brought to a finance or capital planning group to be pared down based on available budgets, historical spending trends, and a number of other subjective factors. An attempt is made to coordinate with broad organizational needs and goals, but this is not done in an objective and structured way. The loudest voices or the best salespeople often get a bigger chunk of the capital resources. Risk reduction, life-cycle costs, return on investment (ROI), and LOS are not formally considered using this approach. A more formal and structured method, represented by the enhanced model in Figure 6-3, provides a utility the analytical parameters that decision-makers seek in today's stakeholder-inclusive and fiscally stringent environment. Using the enhanced model can result in lower overall asset costs and improved results.

The enhanced-model process may appear very time consuming and complex. Although some steps may take a bit more time and effort, the overall effort produces much better results. Organizations report better success in funding acquisition for capital projects, strong consensus on decisions, and lower life-cycle asset costs with approaches similar to those described on the enhanced model.

Operations and Maintenance of Assets

Managing assets to minimize their life-cycle costs while meeting performance requirements is a top priority for utilities and an integral part of effective infrastructure asset management. Effective utilities accomplish such management by employing the correct mix of maintenance techniques at the right point in the asset's operation and life cycle. Many manuals, specifications, standards, and books provide highly detailed information

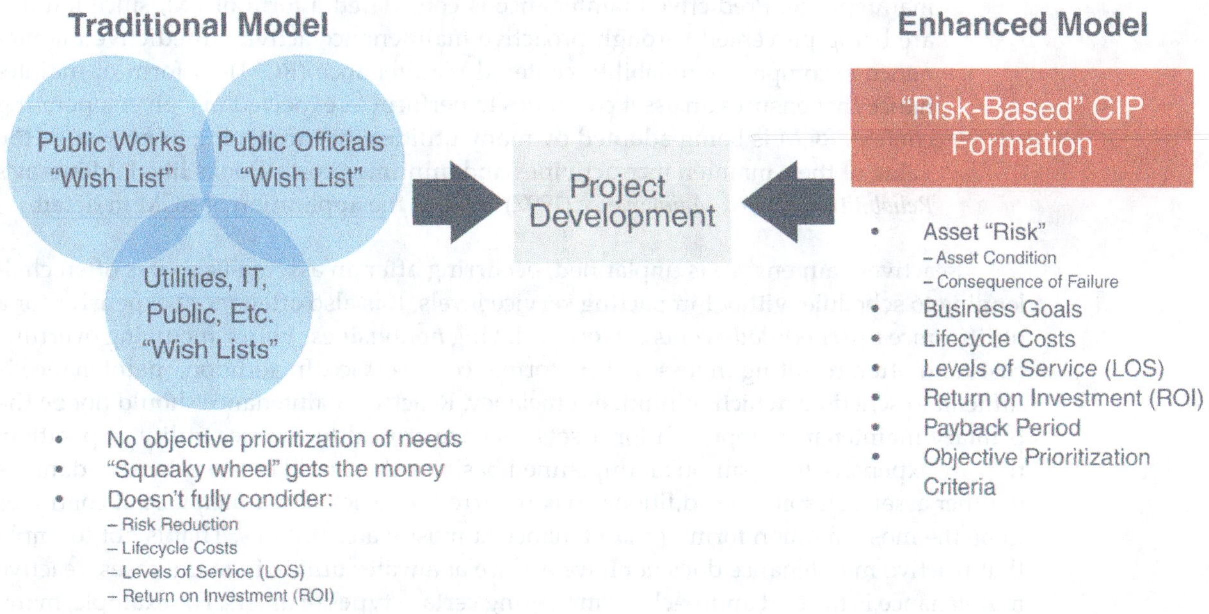

Figure 6-3 **An enhanced model for capital improvement plan (CIP) development offers many benefits over the traditional model**

on asset maintenance practices for different types of assets. AWWA's manuals and specifications are a good place to start for many common water utility assets. An asset's manufacturer is another great starting point to determine what maintenance is recommended and how frequently it should be performed.

Many different types (classifications) of maintenance can be performed on water infrastructure assets. For simplification these practices are broken into three categories:

1. **Reactive or corrective maintenance**—This type of maintenance is performed after an asset failure or performance deficiency is detected. Reactive maintenance includes activities like fixing a water pipe after it breaks, replacing a pump seal in response to a leak, and rewinding a motor when it fails to produce the desired output at the given amperage draw.
2. **Preventative maintenance**—This type of maintenance, which is conducted at regular intervals before an asset failure or performance deficiency occurs, is commonly referred to as *preventative maintenance* (PM). Many utilities that utilize PM reference manufacturers' guidance manuals for suggested PM intervals. Often PM is based on operating hours, on mileage (mobile equipment), or on throughput (flow) measurements. Changing the oil in a car at a given mileage is an example of PM. PM is important for preventing critical asset failures and extending the life of assets in circumstances where maintenance impacts condition.
3. **Predictive maintenance**—Based on observed or calculated data, this type of maintenance, which is conducted in anticipation of a failure or to prevent a performance deficiency, is often referred to as *predictive maintenance*. This approach typically offers cost savings over traditional time-based PM. If skillfully executed, maintenance activities are conducted only when required, thus avoiding overmaintaining an asset, which can occur with PM. Replacing the brakes on a car when the wear indicator begins to make noise is an example of predictive

maintenance. Predictive maintenance is considered a form of PM, since failures are being prevented through proactive maintenance activity. Predictive maintenance encompasses reliability-centered maintenance (RCM)—a form of maintenance that ensures an asset continues to perform as expected in a given operating context. RCM is being adopted by many utilities that are trying to maximize the value of their maintenance activities and minimize costs. Books like J. Moubray's *Reliability-Centered Maintenance* (1997) explain the application of RCM in detail.

Reactive maintenance is unplanned, occurring after an asset fails, so it is often challenging to schedule without impacting service levels. It is also often more expensive for a utility, since unscheduled events can occur during nonbusiness hours, incurring overtime costs and often resulting in losses of performance or service. In addition, maintenance is difficult to schedule, which minimizes efficiency. Reactive maintenance should not be the primary maintenance approach for assets that are critical to a water utility's operation, that are expensive to repair on an unplanned basis, or that could create extensive damage to other assets. Despite the additional risk incurred by reactive maintenance, it continues to be the most common form of maintenance at most water utilities. That is not to imply that reactive maintenance does not have a place at a water utility. In many cases, reactive maintenance is the best approach for managing certain types of assets. For example, many assets with redundant systems or systems the failure of which has little to no impact on customers or on utility operations can be effectively managed with the strategic application of reactive maintenance.

PM and predictive maintenance are planned types of maintenance, which means they are done proactively. They are conducted at predetermined points, in anticipation of a failure or to prevent a performance deficiency. PM is sometimes condition based but more often it is time based. Predictive maintenance is more often condition based. The bulk of a utility's maintenance should be PM and predictive maintenance, leveraging RCM practices. Studies have shown that using the right blend of maintenance types reduces costs and extends the life of assets. The challenge for utility managers is collecting and using the right information to determine which assets should be maintained and at what intervals.

As an example, the different major types of maintenance are represented in Figure 6-4 with an "ideal" percentage that should be represented in the total maintenance picture.

A detailed process for selecting the right maintenance approach for each type of asset is beyond the scope of this text. The basic items to consider in developing that include the following:

- **Review manufacturer's recommended maintenance**—A good place to start for recommended practices for maintaining assets is the manufacturer. Manufacturers can provide baseline guidance on maintenance methods, service and renewal intervals, and common failure modes, but the manufacturer's recommendations are not the end of the story. The manufacturer cannot envision every possible installation or performance situation. Experience, judgment, and an understanding of the operating condition of an asset are important considerations when employing a manufacturer's maintenance recommendations.
- **Understand consequences of failure**—Evaluate the consequences of a failure or performance deficiency associated with an asset on associated equipment, processes, finances, and customers. Understand the relative consequences of failure for all major asset types in a water utility's inventory.
- **Set performance levels**—Establish performance expectations for each asset or group of assets. Determine at which point assets are performing out of their

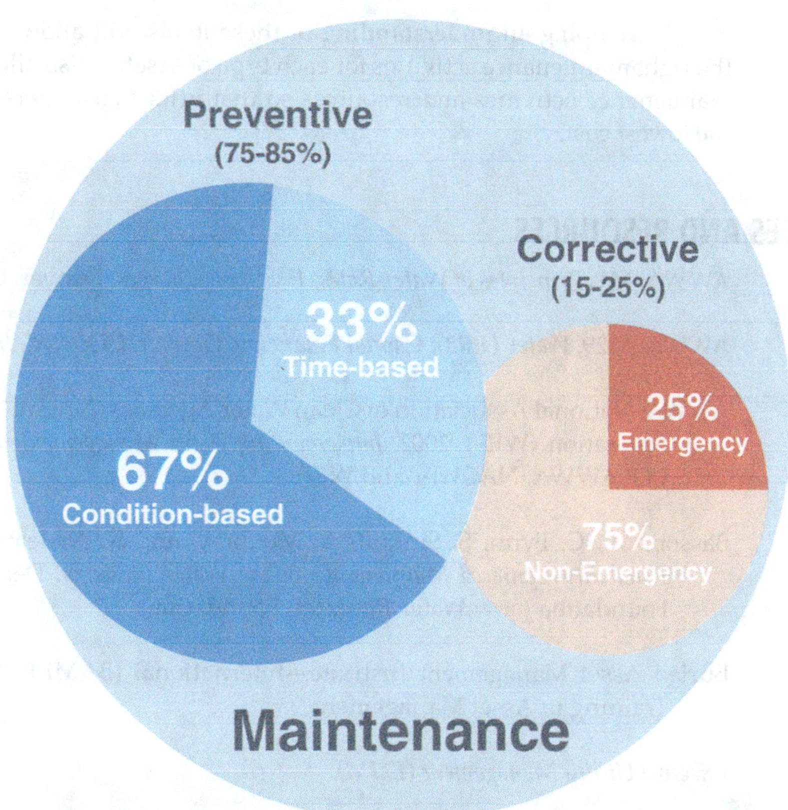

Figure 6-4 Preferred maintenance scenario

acceptable performance window. For example, a pump may have a desired output flow and pressure range that it needs to achieve with a given amperage draw, vibration level, and temperature. Failing to operate in this range, assuming acceptable external operating conditions, is considered an asset performance deficiency.

- **Monitor performance indicators**—Ensure that that the right information is being collected to guarantee that performance levels are achieved. Be careful not to collect unnecessary data that are not needed for understanding the performance of the asset or the impact of maintenance activities. Collecting unnecessary data is wasteful and increases costs.
- **Review failure modes**—Assets fail for many reasons including age-related degradation, random failures, improper installation, faulty designs, natural events, and so on. Reviewing the most common failures modes for each major asset type in a water utility's inventory and designing interventions including maintenance to address them are important.
- **Understand how assets degrade**—Many assets wear out over time and/or with use. Some show little wear over time. A utility manager should review how each major asset type in the manager's system degrades. This assessment should include the relative rate of degradation, degradation indicators, and the expected useful life of assets under different conditions.
- **Take stranded assets into account**—Stranded assets have often had unanticipated or premature shutdowns, devaluations, or conversion into liabilities. An example would be large water treatment plants in times of restricted production caused by drought.

Developing an understanding of these items will allow a utility manager to select the right maintenance activities for each type of asset. It also allows managers to prioritize maintenance activities and resources so that utility customers get the highest benefit at the lowest cost.

REFERENCES AND RESOURCES

AWWA. M1. *Principles of Water Rates, Fees, and Charges*. Denver, CO: AWWA.

AWWA. M29. *Water Utility Capital Financing*. Denver, CO: AWWA.

AWWA, National Association of Clean Water Agencies (NACWA), and Water Environment Federation (WEF). 2007. *Implementing Asset Management: A Practical Guide*. Denver, CO: AWWA, NACWA, and WEF.

Basson, M., C. Fynn, S. Sinkoff, A. Moubray, and R. Nadeau. 2006. *Applicability of the Reliability-Centered Maintenance in the Water Industry*. Denver, CO: Awwa Research Foundation (now Water Research Foundation).

Buried Asset Management Institute—International (BAMI-I). CTAM or Certification of Training in Asset Management.

Effective Utility Management (EUM).

Federal Register (www.federalregister.gov)

Institute of Public Works Engineering (IPWEA). 2015. *International Infrastructure Management Manual (IIMM)*, 5th ed. Australia: IPWEA.

International Water Association (IWA). 2010. *Water Utility Benchmarking: Measurement, Methodologies, and Performance Incentives*. London: IWA.

Lafferty, A. K., and W. Lauer. 2005. *Benchmarking Performance Indicators for Water and Wastewater Utilities: Survey Data and Analyses Report*. Denver, CO: AWWA.

Maticich, M., R. Booth, J. Rogers, E. Rothstein, E. Speranza, C. Stanger, E. Wagner, and P. Gruenwald. 2006. *Asset Management Planning and Reporting Options for Water Utilities*. Denver, CO: Awwa Research Foundation (now the Water Research Foundation).

Moubray, J. 1997. *Reliability-Centered Maintenance*, 2nd ed. New York: Industrial Press.

US Environmental Protection Agency (USEPA). Check Up Program for Small Systems (CUPSS).

US General Accounting Office (GAO). 2004. *Water Infrastructure: Comprehensive Asset Management Has Potential to Help Utilities Better Identify Needs and Plan Future Investments*, GAO-04-461. Washington DC: US General Accounting Office.

US EPA (www.epa.gov)

Water Environment Research Foundation (WERF) and Water Research Foundation (WaterRF). 2012. *SIMPLE—Sustainable Infrastructure Management Program Learning Environment*. Denver, CO: WERF and WaterRF.

WaterID, Water Infrastructure Database. (www.waterid.org). Virginia Tech, Water Environment Research Foundation (WERF), U.S. Environmental Protection Agency, National Science Foundation (NSF).

Additional Resources Available for the New Manager and Team

Key resources related to asset and infrastructure management have been identified throughout this chapter; numerous additonal standards and manuals are available through AWWA for effective management of a utility's assets.

Protection of source water

ANSI AWWA G300—Source Water Protection.

Water treatment plant operations and management

ANSI/AWWA G100—Water Treatment Plant Operation and Management.
ANSI/AWWA B100-B703—Treatment Chemicals.
ANSI/AWWA C110-C153—Ductile Iron Pipe.
ANSI/AWWA C300-C304—Concrete Pipe.
ANSI/AWWA C500-C560—Valves and Hydrants.
ANSI/AWWA C600-C606—Pipe Installation.
ANSI/AWWA C651—Disinfecting Water Mains.
ANSI/AWWA C652—Disinfection of Water Storage Facilities.
ANSI/AWWA C654—Disinfection of Wells.
ANSI/AWWA C900-C950—Plastic Pipe.
ANSI/AWWA D100—Welded Steel Tanks for Water Storage.
ANSI/AWWA D102—Coating Steel Water Storage Tanks.
ANSI/AWWA D103—Factory Coated Bolted Steel Tanks for Water Storage.
ANSI/AWWA D120—Storage.
ANSI/AWWA F101—Contact Molded, Fiberglass Reinforced Plastic Wash Water Troughs.
ANSI/AWWA F102—Matched Die Molded, Fiberglass Reinforced Plastic Weir Plates.
ANSI/NSF 60—Drinking Water Treatment Chemicals—Health Effects.
ANSI/NSF 61—Drinking Water System Components—Health Effects.
AWWA M14—*Recommended Practice for Backflow Prevention and Cross Connection.*
AWWA M17—*Fire Hydrants: Installation, Field Testing, and Maintenance.*
APHA/AWWA/WEF—*Standard Methods for Examination of Water and Wastewater.*

Water distribution system operation and maintenance

ANSI/AWWA G200—Distribution Systems Operation and Maintenance.
ANSI/AWWA B100-B703—Treatment Chemicals.
ANSI/AWWA C110-C153—Ductile Iron Pipe.
ANSI/AWWA C300-C304—Concrete Pipe.
ANSI/AWWA C500-C560—Valves and Hydrants.
ANSI/AWWA C600-C606—Pipe Installation.
ANSI/AWWA C651—Disinfecting Water Mains.
ANSI/AWWA C652—Disinfection of Water Storage Facilities.
ANSI/AWWA C654—Disinfection of Wells.
ANSI/AWWA C900-C950—Plastic Pipe.

ANSI/AWWA D100—Welded Steel Tanks for Water Storage.
ANSI/AWWA D102—Coating Steel Water Storage Tanks.
ANSI/AWWA D103—Factory Coated Bolted Steel Tanks for Water Storage.
ANSI/AWWA D120—Storage.
ANSI/NSF 60—Drinking Water Treatment Chemicals—Health Effects.
ANSI/NSF 61—Drinking Water System Components—Health Effects.
AWWA M6—*Water Meters: Selection, Installation and Maintenance.*
AWWA M14—*Recommended Practice for Backflow Prevention and Cross Connection.*
AWWA M17—*Fire Hydrants: Installation, Field Testing, and Maintenance.*
AWW M28—*Rehabilitation of Water Mains.*
APHA/AWWA/WEF—*Standard Methods for Examination of Water and Wastewater.*

Additional general resources

ANSI/AWWA G400—Utility Management Systems.
AWWA M2—*Instrumentation and Control.*
AWWA M3—*Safety Practices for Water Utilities.*
AWWA M4—*Water Fluoridation Principles and Practices.*
AWWA M9—*Concrete Pressure Pipe.*
AWWA M11—*Steel Pipe: A Guide for Design and Installation.*
AWWA M19—*Emergency Planning for Water Utilities.*
AWWA M20—*Water Chlorination and Chloramination Practice and Principles.*
AWWA M21—*Ground Water.*
AWWA M22—*Sizing Water Service Lines and Meters.*
AWWA M23—*PVC Pipe: Design and Installation.*
AWWA M27—*External Corrosion: Introduction to Chemistry and Control.*
AWWA M28—*Rehabilitation of Water Mains.*
AWWA M31—*Distribution System Requirements for Fire Protection.*
AWWA M33—*Flowmeters in Water Supply.*
AWWA M36—*Water Audits and Loss Control Programs.*
AWWA M41—*Ductile Iron Pipe and Fittings.*
AWWA M42—*Steel Water Storage Tanks.*
AWWA M44—*Distribution Valves: Selection, Installation, Testing and Maintenance.*
AWWA M49—*Butterfly Valves: Torque, Head Loss and Cavitation Analysis.*
AWWA M51—*Air Release, Air/Vacuum Valves and Combination Air Valves.*
AWWA M55—*PE Pipe: Design and Installation.*
AWWA M56—*Nitrification: Prevention and Control in Drinking Water.*
AWWA /ASCE—*Water Treatment Plant Design*, 5th ed; chapters 30, 31 and 32.

Chapter 7

Managing Funding, Operations, and Stakeholders

This chapter describes what water–wastewater managers need to know about the long-term capital-intensive nature of water–wastewater utilities and its impact on relationships with their customers and stakeholders. The purpose is to help managers align planning, funding, and delivery to a common perspective on the cost of doing business.

Water utilities are exceptionally capital intensive, even among utility service providers. This capital intensity means that a relatively large share of annual expenditures is allocated to financing asset acquisition, rehabilitation, and replacement. Because many assets are buried and have extended (and variable) service lives, utility decision-makers face challenges in determining needed levels of reinvestment and securing associated funding. The economics of water is further complicated by the election cycles of public officials and public perception issues, which can significantly affect water agencies' abilities to secure adequate and timely financing for these needed capital investments.

North American water systems have historically not kept pace with the level of rehabilitation and replacement investment required to sustain systems. While improved asset management has become common practice—in part to ensure reinvestments are as effective as practicable—full funding of asset life-cycle costs is atypical. Most utilities face unfunded liabilities associated with their assets. More effective asset management programs are intended to limit and better manage these liabilities.

The long-term capital-intensive nature of water utilities has profound implications for water utilities' relationships with their customers and stakeholders. From a financial perspective, it is important to recognize that the incremental costs to build capacity and provide service extensions to new customers far exceed subsequent incremental costs of delivering that service. By the same token, losses of customer connections are peculiarly significant insofar as they may effectively strand substantial capital investment. The water utility business model is largely founded on supporting economic growth consistent with community values. In so doing, the utility builds a customer base to support annuity payments on a long-term basis.

Significant infrastructure investment is a key driver for customer stability and is largely the reason for a monopolistic operations model. The level of investment makes it impractical to have multiple retail service providers. Other reasons have included the relatively low cost and extraordinary reliability of the services provided and limited recognition of the importance of water resource stewardship. However, these factors are changing with technology advances, new perspectives, and more sustainable water pricing. A variety of conservation and water alternatives (e.g., bottled water, rainwater harvesting, reclaimed/reuse water systems) has already acutely changed water demand patterns and is expected to continue doing so into the future.

The evolving utility environment provides a basis to understand the fundamental irony of water utility economics: Water service is generally priced well below its value, often below its full cost (especially from a long-term perspective), and demonstrably below most customers' willingness and ability to pay. Perhaps the key utility management challenge, therefore, is to ensure resources are in place (through service rates, fees, developer contributions and other charges, grants, public-private partnerships, etc.) that approach levels to sustain reliable service and environmental stewardship.

AWWA recognizes that customers' and the general public's interests are best served through well-maintained and efficiently operated water systems that are sufficiently funded to meet a utility's annual operating and capital needs. In support of this position, AWWA has published and regularly updated the M29 manual, now titled *Water Utility Capital Financing*, to assist utility management and stakeholders in addressing the capital financing needs of their water systems.

The terminology used in M5 is applicable to the United States and Canada; the manual does not attempt to identify practices and terminologies used in the myriad other countries with public utilities. It is likely that many countries outside the United States and Canada have financial instruments and government-sponsored programs that resemble those that are referenced in this manual.

The financial management strategies presented herein are seen by AWWA as universal and thus valuable to all who make use of this document to guide their programs and decisions. The financial operations of many water utilities are continually challenged by a variety of factors, such as adverse weather conditions, inflation, changes in the economy and/or political environment, growth or decline in service demand due to new construction, conservation programs, changes in environmental regulations, backlog of infrastructure repairs and improvements, staff turnover, and, above all, public pressure to keep water rates low. Regardless of such challenges, water utilities remain very capital-intensive businesses. Despite the growing maintenance needs of aging water systems, many water utilities elect to postpone repairs, replacements, and rehabilitation expenditures to keep current operating costs and corresponding service charges as low as possible.

However, deferring capital expenditures invariably results in higher operating costs in the short run and higher construction costs in the future. Strategic planning helps utility management develop the steps necessary to reach the utility's short- and long-term goals and objectives. A strategic financial plan helps management meet the utility's operating and capital financing needs during a specified period. Similarly, strategic capital planning, while consistent with and supported by the strategic financial plan, addresses the utility's smaller recurring capital needs as well as its long-term expansion, upgrade, replacement, and reliability needs.

Strategic financial and capital plans also provide platforms for communicating the utility's needs to the public, thereby gaining greater stakeholder confidence in utility management. The plans inform investors of a utility's potential and support rating agencies' debt ratings. Studies of available alternatives for obtaining funding are a key element in developing successful financing plans. Some financing alternatives come into, or are phased out of, current practice with changes in federal tax law and market conditions. Thus, an effective capital financing program should contain a variety of fundamental approaches to financing infrastructure that create a systematic means of addressing the utility's needs over time. One resource that should not be overlooked is the utility's financial advisor, who is often able to propose current financing alternatives that may provide greater benefit to the utility than originally envisioned in a utility's financing plan.

MANAGING IN AN ERA OF DECLINING SUBSIDIES

Water utilities have been and continue to be subsidized in a number of ways, but the level of subsidy is drastically declining in parallel with increasing infrastructure and operating costs. Subsidies have historically included federal and state grant programs and low-interest loan programs. These programs have made water infrastructure funds available under exceptionally favorable terms. Available funding has declined dramatically over the past two decades, but still remains a pillar of infrastructure financing. Because most water utilities in North America are public entities, they have access to tax-free municipal debt in addition to low-interest loan programs. These sources provide an additional form of subsidy. Recent advocacy for these and alternative forms of subsidy, including the concept of a federally sponsored Water Infrastructure Finance Innovation Authority (WIFIA), suggests that industry subsidization will remain an important attribute of the industry's financial landscape. Uncertainties about the extent and reliability of external subsidies necessitate that utility managers develop strategic financial plans that consider a number of different funding options.

The Drinking Water State Revolving Fund (DWSRF) program is a federal–state partnership to help ensure safe drinking water. Additionally, there is the Clean Water State Revolving Fund (CWSRF). The CWSRF was established by the 1987 amendments to the Clean Water Act (CWA) as a financial assistance program for a wide range of water infrastructure projects, under 33 US Code §1383. The program is a powerful partnership between USEPA and the states that replaced USEPA's Construction Grants Program. States have the flexibility to fund a range of projects that address their highest-priority water quality needs. The program was amended in 2014 by the Water Resources Reform and Development Act.

Using a combination of federal and state funds, state CWSRF programs provide loans to eligible recipients to

- Construct municipal wastewater facilities
- Control nonpoint sources of pollution
- Build decentralized wastewater treatment systems
- Create green infrastructure projects
- Protect estuaries
- Fund other water quality projects

Building on a federal investment of more than $39 billion, the state CWSRFs have provided more than $111 billion to communities through 2015. States have provided more than 36,100 low-interest loans to protect public health, protect valuable aquatic resources, and meet environmental standards benefiting hundreds of millions of people.

STRATEGIC FINANCIAL PLANNING

By employing a strategic financial plan, a utility can define the revenue requirements to meet current and future service obligations; identify gaps between available funding and infrastructure development, rehabilitation, and replacement needs; and evaluate prospective utility system financial performance and financial integrity. A strategic financial plan includes financial policies to guide a utility's financial development, specifying debt service coverage targets, levels and uses of reserves to meet unexpected costs, and targeted capital structure attributes (e.g., debt/equity ratio).

Conceptually, strategic financial planning is a relatively simple exercise, generally amounting to the development of multiyear cash flow and debt analyses. These analyses are typically for five or ten years but may be longer for those utilities contemplating major long-term capital investment programs. A strategic financial planning model enables revenue projections under alternative assumptions related to customer account growth, billable usage, interest earning rates, varying weather and usage patterns, and other factors. Similarly, operations and maintenance expenses may be projected under alternative assumptions related to staffing, inflation of key expense items, treatment requirements, and implementation of new technologies. Finally, cash flow modeling may define the utility's optimal capital financing strategy by testing implications of the use of different forms of debt obligations (e.g., revenue bonds, loans), use of equity financing, and application of dedicated funds (e.g., reserves).

RATES/COST-OF-SERVICE ANALYSIS

While strategic financial planning will provide insight into annual systemwide revenue requirements, specific rates for individual customer classes—if adjustments are not implemented uniformly or across the board—might be determined through a cost-of-service analysis (COSA). Authoritative guidance for the procedures that may be used for this form of analysis is provided by AWWA M1, *Principles of Water Rates, Fees and Charges.* COSA is used to determine an equitable distribution of revenue responsibilities across customer classes by allocating costs on the basis of cost causation. COSA distributes costs to customer classes in proportion to those classes' shares of different demands placed on the water system. For example, if residential users (as a class) impose 50 percent of average day demands and 75 percent of peak day demands, and represent 90 percent of the accounts served, this class may be assigned revenue recovery responsibility for 50 percent of average day demand related costs, 75 percent of peak day related costs, and 90 percent of customer account related costs.

COSA is fundamentally a three-step process, although the second step typically involves a three-step subprocess as illustrated in Figure 7-1. The overarching steps are (1) determination of rate revenue requirements, (2) cost allocation, and (3) rate design. Within the cost allocation step, costs are initially assigned to utility functions, then to service characteristics, and ultimately to customer classes.

Revenue Requirements

Revenue requirements are the utility costs that must be recovered through various sources. Total revenue requirements are determined by assessing the costs that are necessary to operate and maintain the utility and to finance capital investments for system expansion, upgrades, and renewal, as well as to maintain adequate liquidity in the event of unplanned expenses or revenue shortfalls. Recovering appropriate revenue ensures financial sustainability. Identifying the appropriate amount of revenue requirements is an important first step in establishing user charges.

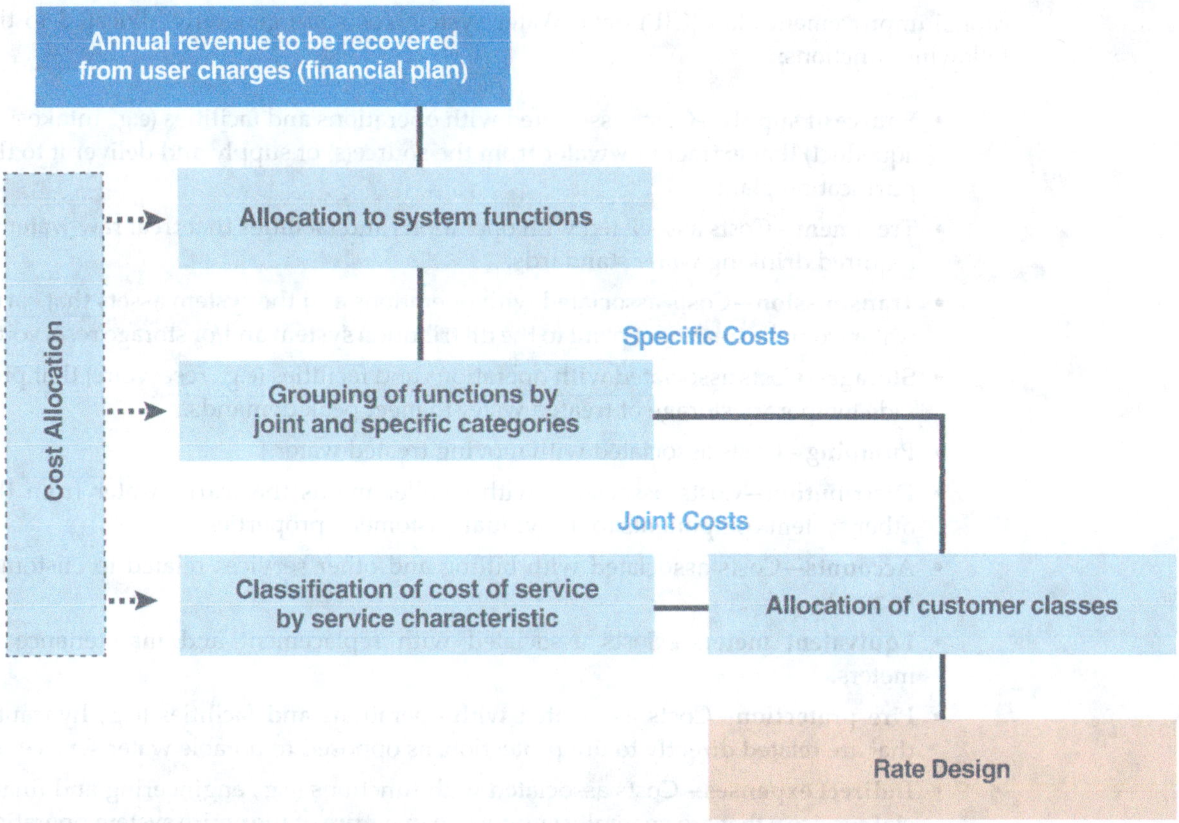

Figure 7-1 Three-step COSA process

Revenue requirements are usually divided into two major categories: (1) operations and maintenance (O&M) costs and (2) capital costs. O&M costs are usually routine or periodic costs incurred by a water utility in providing service. Capital costs relate to capital items such as equipment or facilities that provide benefits over multiple years.

The two basic approaches for establishing revenue requirements are (1) the cash needs approach and (2) the utility approach. Generally speaking, the cash needs approach focuses on recovering the annual cash needs of the utility, whereas the utility approach is more directly tied to accounting principles and provides for a rate of return on investment.

Cost Allocation

The basic steps used to allocate the revenue requirements of the water systems to customer classes include (1) categorization of revenue requirements by utility function (functionalization), (2) classification of functionalized costs based on service characteristics (classification), and (3) allocation of costs by service characteristic in proportion to the use of each customer class (allocation).

UTILITY BUDGETING

Water utilities incur costs to perform different system functions needed to meet customer service requirements. Operating costs are typically allocated to functions based on budget categories that capture different functional distinctions and/or evaluation of operating activities (e.g., work orders). Capital costs are generally divided based on fixed asset and capital improvement plan (CIP) data. Water system costs are generally allocated to the following functions:

- **Source of supply**—Costs associated with operations and facilities (e.g., intake and aqueduct) that extract raw water from the source(s) of supply and deliver it to the purification plant.
- **Treatment**—Costs associated with operations and facilities that treat raw water to required drinking water standards.
- **Transmission**—Costs associated with operations and the system assets that carry water from the treatment plant to the distribution system and/or storage reservoirs.
- **Storage**—Costs associated with operations and facilities (e.g., reservoirs) that provide temporary storage of treated water to meet peak demands.
- **Pumping**—Costs associated with moving treated water.
- **Distribution**—Costs associated with smaller mains that carry water from the other system components to individual customers' properties.
- **Accounts**—Costs associated with billing and other services related to customer accounts.
- **Equivalent meters**—Costs associated with replacement and maintenance of meters.
- **Fire protection**—Costs associated with operations and facilities (e.g., hydrants) that are related directly to fire protection, as opposed to potable water service.
- **Indirect expenses**—Costs associated with functions (e.g., engineering and financial services) that are generally supportive in nature of the entire system operation or a subset (e.g., administration within a specific division). These costs are generally reallocated back to the other functions in proportion to the directly allocated operating or capital costs.

Classification to Service Characteristics

Some costs incurred by a utility are a function of the quantity of water consumed during average demand and peak demand conditions. These are referred to as *variable costs*. Other costs are associated with serving customers regardless of the quantity that flows through the system. These are called *fixed costs*. Costs are further classified by the following characteristics so that they may then be allocated:

- **Base or average day** costs include capital costs and O&M expenses associated with service to customers under average demand conditions, but do not include any costs attributable to variations in water use resulting from peaks in demand. Base costs tend to vary with the total quantity of water used.
- **Maximum-day and maximum-hour** costs include costs attributable to facilities that are designed to meet peaking requirements. Such costs include capital and operating charges for additional plant and system capacity beyond that required for average rate of use.
- **Customer or billing** costs are associated with any aspect of customer service, including billing and accounting services. These costs are independent of the amount of water used and the size of the customer's meter and are not subject to peaking factors.
- **Equivalent meter** costs are associated with the purchase and maintenance of water meters. Equivalent meter factors equate the hydraulic capacity of meters of

different sizes to calculate the total number of residential-equivalent meters (5/8-in. to 3/4-in. meters) connected to the system in each customer class.

- **Fire protection** costs include the costs of providing fire protection to the community.

O&M costs are generally allocated based on the operation and design of facilities, while capital costs often are classified based on facility design criteria.

Allocations to Customer Classes

Allocation of costs by service characteristic to customer classes is relatively straightforward once costs have been determined for individual service characteristics. These costs provide the basis for a unit cost calculation. For example, projected average demand-related costs (determined through classification) are divided by projected average demands. Cost responsibility by customer class may then be determined by multiplying the unit costs by the projected units of service for each customer class. Cost-of-service based rates are defined as costs that fully recover these fully allocated costs by customer class.

Rate Design

While cost-of-service rates are those that recover fully allocated cost responsibility by customer class, there are literally hundreds of rate structures that may accomplish this objective. Rate design is the process by which specific rates and charges for service are determined. The structure or design of a utility rate refers to the set of unit charges and customer use measures used to calculate bills for individual users. In most cases, utility rates include fixed charges that do not vary with customer usage, often called *customer charges* or *service charges,* and one or more charges per unit of demand (however that demand is measured). In some instances, a quantity allowance is included in the customer charge; in other cases, charges are assessed on a customer's total use.

The art of ratemaking involves designing rates that balance inherently conflicting rate objectives in a manner that reflects community values. At a minimum, utility rates should be sufficient to generate revenues required to support operations, maintain and develop capital infrastructure, and preserve (or enhance) the financial integrity of the utility system. In addition, there are a number of other technical and policy objectives that utilities often set out to achieve in developing rate structures. These may include rate stability, equity, simplicity, and public understanding. Utilities may seek to support or promote resource efficiency and/or economic development through their rate designs, and many communities are increasingly implementing measures to ensure the affordability of service for low-income ratepayers.

System development charges. An additional mechanism to capture costs is the use of system development charges (SDCs), often referred to as *impact fees.* These fees have significant policy ramifications and encompass a broad array of fees imposed for connection to a water system. These fees are generally oriented toward recovery of growth-related capacity costs from new system users. Notably, these fees have been the subject of considerable scrutiny and litigation as they reflect a utility's support of community growth management policies. In most states, procedural and methodological requirements regarding SDCs have been established through legislation and/or case law.

Miscellaneous charges. Most utilities recover a variety of costs for services provided to limited numbers of customers through a variety of direct miscellaneous charges including site plan review fees, late payment charges, or service turn-on/turn-off fees. These fees should reflect the costs required for delivery of these specific services.

OPPORTUNITIES FOR EFFICIENCY

The principles for development of appropriate and defensible rates, fees, and charges are predicated on the notion that water system costs are reasonable and consistent with requirements to deliver reliable services. As the infrastructure-funding gap has come to the forefront of policy discussion, the desire to achieve efficiency has become more acute. North American water utilities have historically had a mixed record of accomplishment, and significant opportunities for efficiency gains remain palpable, as noted in the following sections.

Institutional Structures and Consolidation

Perhaps the most noteworthy opportunities for efficiency gains derive from re-evaluation of institutional structures that govern the sector. In contrast to many other parts of the world, North America is populated by thousands of utility providers with different organizational structures. The excessive number of utilities often limits the extent to which economies of scale may be achieved, precipitates undue redundancy in operational capacity, and invites (often unproductive) political complexities. Recognizing this situation, many utilities have voluntarily sought to collaborate on aspects of service delivery ranging from development of regional treatment facilities to coordination of security measures. Barriers to structural consolidation and regionalization reflect local government complexities that historically have proven to be exceptionally difficult to overcome (especially with relatively low service costs). As costs continue to rise, these barriers are likely to become more surmountable.

Operational

Beyond institutional reform, water utilities throughout North America achieved notable operating efficiency improvements (prompted in part by real or perceived threats of privatization) in the late 1990s and 2000s, and they continue to focus on opportunities to gain efficiency. These improvements in efficiency derive from a host of aspects of utility operations and may differ markedly across individual utilities. Some of the most significant include but are not limited to the following:

Treatment process optimization. Operating savings may be achieved through a variety of means from the selection and configuration of treatment processes to disciplined chemical dosing.

Energy management. Water utilities are among the most significant users of electricity in North America. Cost reductions may be achieved through, for example, installation and use of high-efficiency equipment (e.g., variable-speed-drive pumps) and the timing of power use to take advantage of off-peak electric rates.

Strategic staffing. Staff reductions, often without requiring involuntary reductions in force, may be achieved through business process optimization analyses in combination with effective succession planning and managed attrition.

Work order management/technology enhanced field operations. Among the most significant business process optimization techniques is the deployment of various technology solutions. Work order management systems help utility managers ensure field crews are dispatched with appropriate equipment and supplies to locations in ways that limit unproductive time. Field technologies (e.g., "tough book" Toughbook® field computers) facilitate documentation of conditions found, work performed, and resources expended at project sites.

Meter-reading technologies. Various meter-reading technologies have limited the extent to which human meter readers are needed to collect billing data. Automated metering infrastructure (AMI) and advanced meter reading (AMR) also can facilitate more efficient identification of sources of water loss and metering problems (as well as enhance revenue forecasting) through high-resolution monitoring of customer demand patterns.

Capital Project Delivery

As new technologies and business processes have yielded important operating efficiencies, similar gains have been and continue to be realized in delivery of water system capital infrastructure. While some efficiency gains have been accomplished through technology enhancements to traditional design-bid-build processes, various alternative project delivery (APD) mechanisms have yielded profound opportunities for cost and schedule reductions. These mechanisms typically involve integration of one or more traditionally distinct project phases including design, construction, operations, and financing.

The selection of project delivery methods turns on risk allocations that highlight the challenges of managing large infrastructure investments. Capital project delivery risks include but are not limited to:

- Design quality
- Timing (schedule)
- Site conditions
- Changes in demand
- Supplies and materials cost escalation
- Construction performance
- Financing terms (interest rates)

Traditional design-bid-build approaches enable utilities to have substantial control over each phase of project delivery—ensuring conformance to utility-prescribed practices and procedures—and acceptance of most forms of project delivery risk. Integrated approaches require ceding of some control, with associated risk transfers, to obtain cost and schedule advantages.

SECURING STAKEHOLDER SUPPORT—"SILENT SERVICE" NO MORE

Decisions related to project delivery processes as well as many (if not most) other aspects of utility management are subject to scrutiny by interested stakeholders, particularly for public water systems. Whereas water utility operations and system development may have garnered relatively less interest and attention in the past because of relatively low rates and less access to information, twenty-first-century water utility managers should expect to involve interested stakeholders in key decisions and be held accountable by these same parties. Water utility managers have to recognize that the days of water utilities anonymously providing reliable service at low costs in relative silence are in the past. See chapter 8 for additional discussion of stakeholder engagement and communication.

Demands for stakeholder engagement derive from a multitude of concerns that have grown more pronounced, including but not limited to

- **Public health**—Water utilities are charged with providing reliable water services that meet treatment standards to protect public health. As individuals and community groups focus on questions related to potable water quality, the relative advantages and disadvantages of treatment technology options, and the merits of emerging regulations, water utility management will be called on to both delineate its performance and procedures and provide expert insight.
- **Environmental stewardship/water supply and conservation**—Water utilities are charged with ensuring that adequate water supplies will be in place to serve community demands, and are increasingly called on to ensure efficient, sustainable management of water resources. Water utility management must provide reassurance of adequate supplies and effective environmental stewardship.
- **Capital project/construction planning**—Perhaps more fundamentally, the age of unquestioned system development where utilities independently decided about the routing, configuration, and aesthetics of public infrastructure has waned. Interested stakeholders have come to expect to participate in capital projects and construction scheduling. Water utility managers must design and implement effective stakeholder engagement initiatives to move projects forward.
- **Rate and fee setting/affordability**—As water utility rates and fees continue to increase, water utility managers will be called on to demonstrate not only efficiency but also the equitable distribution of cost responsibilities—primarily through strategic financial planning and COSA as discussed previously. Stakeholder engagement in these analyses is becoming standard practice, as are studies that evaluate system development charges/impact fees as components of typically highly scrutinized growth management policies. As water rates increase, concerns related to the affordability of service for low-income populations are becoming more acute and are increasingly compelling utilities to consider sponsoring or implementing programs designed to ensure that these populations continue to have access to services needed to protect public health.

Fundamentally, water utility customers are informed by, and derive their expectations from, an entirely different customer service environment than that which characterized utility operations throughout much of the last century. Water utility customers are also telephone utility customers who now are served by mobile phones that function as small personal computers. They also participate in community groups that are engaged in all aspects of local government planning and development. Water utilities that engage their communities, embrace innovation and deploy technological advances, and consistently demonstrate efficiency, effectiveness, and equity in their day-to-day efforts, are well positioned to achieve balance and success in managing their operations, acquiring funding, and garnering stakeholder support.

REFERENCES

AWWA. Manual M1. *Principles of Water Rates, Fees and Charges*. Denver, CO: AWWA.

Chapter 8

The Issue: Utility Communication and Stakeholder Engagement

This chapter contains a description of steps utility managers can take to improve relationships with customers using various tested communication approaches. These steps can be adopted by utility managers as the foundation for dealing with challenges in residential, commercial, and corporate water–wastewater customers. Moreover, it is necessary to connect communication actions with the strategies and methods in each of the other subjects discussed throughout this manual.

Public outreach and communication with stakeholders and customers are key management functions in a utility. They are not "nice to do" activities if time and budget allow. They are as critical to a smoothly functioning utility as capital planning, system maintenance, human resources, rate planning, and the many other basic ingredients of utility management.

An effective communication approach fundamentally establishes a link between the proposed action and the best interests (short term or long term) of the stakeholders. Implicitly, stakeholder interests are both operational and financial. For instance, rate increases are often (but not always) in the best interests of the ratepayers from an operational standpoint—they allow the utility to complete projects to keep the water safe for consumption or prevent service interruption.

Identify a utility that continues to earn support for essential capital improvements, rate increases, and utility policy decisions and you will undoubtedly also find that utility has a strong, organized, and strategic communication planning program—and that the program is implemented consistently over time. Strategic communication planning is a key component of a utility's efforts to proactively address myriad opportunities and challenges, and management support is *the* key ingredient to the communication program's viability.

In utilities where management views public outreach as a critical part of strategic planning and day-to-day utility operations, the public outreach program is successful. On the one hand, successful communication programs equal

- Increased community collaboration for projects and utility programs
- Public support for decisions by policy makers
- Increased recognition among community members of the value of water and its role in healthy, vibrant communities
- Fewer concerns expressed about rate increases and other potentially controversial issues

On the other hand, many utility professionals can point to challenges they have faced because of inaccurate or misleading information, including

- Failed or delayed rate increases
- Delayed construction projects and management proposals
- Public accusations of mismanagement
- General concerns and misunderstandings among utility stakeholders

These challenges may become more frequent and magnified as the water utility faces costly and expansive infrastructure improvements, competition for resources, and double-digit rate increases to support critical services, not to mention the national focus on security issues and water quality.

WHAT UTILITY MANAGERS CAN DO TO ESTABLISH A SUCCESSFUL OUTREACH PROGRAM AND IMPROVE RELATIONS WITH THEIR COMMUNITIES

Develop a plan. A plan will take effort and resources but will result in a thoughtful, tailored strategic path forward and a basis on which to prepare annual budgets associated with outreach and engagement. Just as utilities have guidance documents for their operations or capital improvement programs, they also need a planned path forward for public outreach. Written, structured, methodical, and strategic communication plans should guide public outreach efforts.

Plans include

- **Situation analysis**—A description of the situation that is the subject of the communication or public outreach.
- **Program challenges and opportunities**—A summary of potential issue challenges and opportunities.
- **Goal**— Why does your utility want to communicate about this issue or program?
- **Objectives**—The results you want to achieve.
- **Messages**—What you want to say about the issue and what you hope your audience will take away as a result of the communication.
- **Audiences**—Who you want to communicate with.
- **Strategies**—How the communication goal and any objectives will be accomplished.

- **Tactics/activities**—Specific activities that you will conduct or tools you will use to accomplish your objectives.
- **Evaluation**—Measuring the effectiveness of the tactics and activities that you have employed and course correcting when needed.
- **Surveys**—Regularly occurring customer and stakeholder satisfaction inquiries, with measurable and comparable outcomes.
- **Timeline for implementation**—The schedule you need to follow (closely tied to the organizational or project milestones that your communication plan is supporting).
- **Roles and responsibilities**—Designation of communication team members and assignment of those who will implement your plan.
- **Rapid response planning**—Considering the potential for emergency or crisis situations and establishing a protocol that will enable fast-paced but accurate communication to customers and the media.

Make sure communication professionals are part of all utility planning teams. Public outreach does not stand alone any more than any other component of your utility. All utility planning requires a dedicated team of professionals working together—resource planners, environmental analysts, financial experts, and engineers often comprise this team. But often, a critical piece of the project team is missing and should be added: the communication professional. If you are communicating with the right people in your community, in the right way, at the right time, your potential for project success without delays is exponentially higher.

If you do not have a communication professional, find one. Because not all utilities have dedicated communication professionals, sometimes managers will need to think outside the box to identify skill sets among utility personnel. Contract services are also available to help prepare and implement outreach plans.

Put a face with a utility name. You want your customers and community members to recognize more than your logo. One or more trained utility spokespersons should be conducting interviews, making presentations, and participating in community events. This consistent interaction not only provides important information and allows for two-way communication but also demonstrates that the utility is far more than the bill customers receive in the mail and forms the basis of establishing trust and credibility.

Keep in mind, there is no such thing as a born communicator. Targeted written communication requires skill. Speaking before stakeholder groups requires an ability to speak in terms that can be understood by the layperson, avoiding the jargon and acronyms rampant in the water profession. And during emergencies, the ability to communicate with empathy will be significant. These and other skills can be honed and maintained through continuous practice and training scheduled with the support of the utility manager.

Invite the community to help and weigh in. Asking for community input on basic utility functions can be a tough undertaking for many utility professionals. But the benefits to the utility that has partnered with its community can be critical to implementing a successful program or project. Working with community members can help ensure they understand your operations or proposals and allows stakeholders to point out potential bumps in the road before proposals are too far along.

Use the input you asked for. Another way to build community support is to actually use the input you asked for and demonstrate how you have used it. To maintain openness and transparency, utilities also need to explain why they are not able to implement some of the recommendations received.

Throw away the cookie cutter. Your utility outreach needs to fit your community. One of the most important tasks utility managers have is to understand their community and its character, history, values, and culture. What is acceptable and desired in one community might be quite different in another. The same is true for the ways in which community members desire to receive information, or the messages that resonate best with community members. Where social media may be appropriate for one community, door hangers or fliers at the grocery store may be appropriate in another. Know your community.

Often, effective communication results from explaining to the public the operational or financial risks of *not* taking action. For example, generally a utility's capital needs are not anticipated to go away but, due to historical inflation in the cost of construction, may be more expensive to address in the future. As such, user rates could be even higher in the future if no action is taken. Also, during periods in which interest rates are low, there could be interest rate risk if the utility would like to consider borrowing to fund a portion of its capital program. For regulatory or compliance issues, there is the risk of fines or consent orders. Operationally, management can often outline the risks of system failure in certain parts of the service area if no action is taken.

Ensure communication is constant and sustained. Maintaining ongoing communication with utility stakeholders is ultimately the responsibility of the utility manager. Many resources are available to help establish and maintain effective interactions and communication with community members, elected officials, customers, and other stakeholders. The basics of good utility communication are accessible in AWWA documents and programs that can be found through awwa.org.

At times a manager must engage in more technical or complex communication as a result of a consent decree or other requirement from another agency. Some case examples of this type of communication are outlined later in this chapter.

Summary

The success or failure of the new water utility manager will largely depend on how well that water utility manager has communicated. When a utility is weathering the regulatory, legal, or financial storm of the day, a good communications strategy can provide needed shelter from the storm.

To best manage various communication challenges that regulatory compliance creates, utilities need to define, articulate, and share their communication strategy internally (communication plan) and externally, and the plan needs to closely align with regulatory compliance obligations. A communication strategy can enhance a utility's ability to comply with statutory and regulatory requirements in many ways, including

- Making public the utility's statutory and regulatory accomplishments
- Documenting performance measured by third parties (regulatory and financial)
- Explaining compliance issues trends and their impacts
- Maintaining access to data for audit purposes
- Using maps and graphic user interfaces for accountability and ease of communication
- Demonstrating ability to manage and leverage the risks and advantages of mobile computing, the proliferation of social media, cloud computing, connectivity, interoperability, and the growth of business intelligence
- Making the information accessible to the consumer
- Providing a proactive message on data security

Communication plan. Therefore, the development of a communication plan should address the following business needs:

- Aligning communication technology with the stated mission, vision, goals, and objectives
- Defining a strategic direction and plan for all communications
- Implementing best practices for managing communication, taking advantage of existing investments in information technology systems and professional networks
- Identifying key organizational and staffing events planned to support communication goals

A successful communication plan must match a utility's defined business objectives and gain the buy-in of the entire organization including management, operations, maintenance, finance, customer service, laboratory, and engineering personnel. Achieving improved organizational efficiency within a diversified and complex organization requires meaningful integration of communication and business practices throughout the organization. The resulting communication plan should also identify and eliminate redundant and inefficient information silos such as orphan databases.

In designing the communication plan to address the utility's most pressing business needs, the first step is to include a thorough understanding of customer needs and related business-process initiatives on the departmental level, as well as any competing citywide or enterprisewide communication initiatives.

The second step in developing a communication plan is to inventory and assess the current communication efforts of associated business processes. The next step is to develop applicable strategies for reconfiguring, enhancing, or potentially replacing existing communication approaches. Once these are identified and prioritized as a series of defined projects with budget, schedule, and resource needs, planning can take place for the next immediate period (< 2 months), short term (several months), medium term (up to a year), and/or long-term (multiple years) horizon.

CASE STUDIES

As noted previously, varied outreach methods are available to meet specific communication and stakeholder goals, and each must be strategically tailored to meet the needs of the utility and its community. The following two actual case examples describe circumstances commonly faced by many water utilities and communication methods to meet utility objectives by implementing effective outreach and stakeholder involvement.

Example 1: Rate increase.
Situation.
A water utility in the western United States had relied on groundwater to serve customers, but that water supply source was decreasing. To serve existing customers and new customers would require conversion to surface water. Converting to a new water supply source would require significant funding and require increasing user rates and development fees. Based on community opposition, the utility board declined to raise rates and fees until additional actions were taken, including involving the public in a comprehensive way.

Communication strategy.

Generate long-term stakeholder support for necessary capital projects to ensure a reliable water supply through community engagement and through providing consistent, transparent information to the community.

Tactics employed.

- Identified overarching communication objectives.
- Developed key messages and tested messages through surveys and one-on-one interviews with community leaders.
- Updated key messages to include in informational materials and conducted staff workshops to increase familiarity with messages.
- Developed web site content.
- Created and designed information materials, including a water supply management plan brochure, fact sheet, and list of frequently asked questions (FAQs).
- Drafted news releases about the rate study and key milestone events.
- Developed and designed an advertorial campaign highlighting important water issues.
- Convened and facilitated a community working group to assist staff in their review of the rate study and provide community input to the consultant and staff regarding recommendations in that study.
- Organized and conducted community tours of utility facilities.
- Incorporated measures to evaluate audience reach and communication effectiveness.

Results.

The community working group, representing a wide variety of stakeholders, played an important part in the discussion and saw its input reflected in the rate study report, as promised. The report led to a board vote that raised rates based on clear, factual information, without public opposition. The variety and focus of communication material measurably affected customer understanding, leading to a rejection of a recall election aimed at board members who studied and supported the critical increases.

Example 2: Construction relations.

Situation.

A water utility was faced with a long-term, large-scale infrastructure improvement program to replace aging pipelines through the city. Spanning more than 80 individual projects, the construction program would create long-lasting improvements and upgrades to the water system infrastructure but would also affect community members and elected officials for many years through increased rates as well as traffic and aesthetic impacts. Environmental justice concerns were also raised based on the location of many construction sites.

Communication strategy.

Raise and maintain broad awareness of the importance of the capital improvements, the long-term benefits to the community, and the city's efforts to minimize construction impacts and inconvenience.

Tactics employed.

- Developed a programmatic communication plan and associated staffing plan to ensure outreach support for the many construction projects.
- Developed key messages to include in the communication strategy and informational materials.

- Provided routine training to ensure consistent staff knowledge of the program benefits and ensure the agency spoke with one voice to all stakeholders.
- Conducted presentation, media relations, and crisis communication workshops for project managers and engineers.
- Developed a library of informational materials including fact sheets, FAQs, pocket cards for field crew distribution, kiosks and posters, and fliers and door hangers for targeted delivery.
- Developed an interactive web site to provide in-depth information and accept public questions and inquiries.
- Launched a variety of social media platforms to provide project updates and real-time information about construction impacts such as road closures and related traffic inconveniences .
- Expanded social media tools such as blogs, Facebook, and Twitter to communicate information about individual projects and activities.
- Conducted community meetings, special events, and media tours.

Results.

Although many individual projects were moving forward simultaneously and in different geographic areas, a programmatic approach to stakeholder outreach helped create a strategic road map of activities that not only organized and focused efforts of the program team, but also fostered a clear understanding and awareness among external audiences about the importance of this critical capital improvement program.

REFERENCES FOR DOCUMENTS AND PROGRAMS AT AWWA AND ELSEWHERE

AWWA's web site, awwa.org, has a wealth of information about communication and public outreach for utility managers. An associated web site that provides examples of communication materials is drinktap.org. In addition, AWWA recently published the second edition of *Thinking Outside the Bill,* which is a quick reference developed exclusively for AWWA member utilities. *Journal AWWA* has included articles about public outreach and communication, such as "Examining the Benefits of Effective Public Outreach" published in the January 2012 edition (Vol. 104, No. 1), which includes perspectives from a utility manager, an engineer, and a public outreach practitioner. *Opflow* also includes articles about specific projects that demonstrate how public outreach has been incorporated.

Resources found on awwa.org will be particularly helpful to utility managers. Once on the web site, click on Resources & Tools and then click on Public Affairs to access the following:

- Public Information takes the user to several useful tools including free resources such as The Drip Calculator, a variety of publications, and drinktap.org, a Web site dedicated to water quality issues.
- Communication Tools provides a wealth of information for water utility managers, including the excellent information program Only Tap Water Delivers that includes ads, presentations, and more for utility members to personalize and use.
- Public Communications Toolkit—leads to a wide variety of communication resources that are categorized by Strategic Planning, Media Relations, Events, Crisis & Issue Communications, Community & Youth Outreach, and Digital & Electronic Communications. Within each of these categories the user will find links to articles, guides, and tips; best practices for consumer confidence reports;

specific-issue information such as conservation, infrastructure, and pharmaceuticals; and more.

- Under the Strategic Planning subheading in the Public Communications Toolkit, the user will find that AWWA's G420 standard, Communication and Customer Relations, is just a click away.

All utility managers are encouraged to review the array of information and guidance documents pertaining to communication and public outreach available on awwa.org as a supplement to this manual.

Chapter 9

Environmental Health and Safety

The water utility industry is charged with providing for public health (drinking water) and environmental quality (wastewater treatment, watershed management). To meet those obligations, utility managers must be knowledgeable about and fully compliant with a variety of national, state or provincial, and local laws and regulations. Some states and provinces may establish higher standards than those set at the federal level. More important, because of the nature of the work utilities perform, they must establish strong health and safety programs to protect employees and the general public. This chapter reviews some of the key points managers must address in these areas. Compliance is less costly than accidents, injuries, and fines and causes less stress on the organization and its resources.

From its inception, AWWA has served as a public health organization by setting voluntary standards, soliciting public comments on all proposed new AWWA standards, and doing the same on all proposed revisions, reaffirmations, or withdrawals of existing AWWA standards. AWWA creates standards addressing the entire industry spectrum of treatment plant operators and managers, scientists, environmentalists, manufacturers, academicians, regulators, and utilities. Most recently, ANSI/AWWA C810, Replacement and Flushing of Lead Service Lines, has drawn national attention. C810 was approved, published, and became effective in 2017. This exemplary standard describes essential procedures for the replacement of lead water service lines and flushing following replacement. Essential procedures include appropriate tools and techniques; flushing a service line after replacement; factors to consider in optimizing flushing; and instructions to provide customers affected by the replacement including additional risk reduction measures. ANSI/AWWA C810 also describes procedures for partial replacement and repair situations where complete lead service line replacement is not possible or practical. This standard is one of many to consider for environmental health and safety.

Equally important to environmental health are the collection and treatment of domestic sewage and wastewater. As stated by USEPA, collection and treatment are *among the most important factors responsible for the general level of good health enjoyed in the United States*. Sewers collect sewage and wastewater from homes, businesses, and industries and deliver it to wastewater treatment facilities before it is discharged to water bodies or land or reused.

National Pollutant Discharge Elimination System (NPDES) permits establish discharge limits and conditions for discharges from municipal wastewater treatment facilities to waters of the United States. Public health requires overview of processes used to treat domestic wastewater before discharge to the nation's waters. This is largely done through the NPDES Permitting Framework, which is a framework for establishing water quality and technology-based NPDES permit limits. There are also secondary treatment standards, which are minimum, technology-based standards for discharges from municipal wastewater treatment facilities.

Public health regulations cover collection systems, and generally municipalities have used two major types of sewer systems: combined sewers and separate sanitary sewers.

- **Combined sewers** are designed to collect both sanitary sewage and stormwater runoff in a single-pipe system. These systems were designed to convey sewage and wastewater to a treatment plant during dry weather. Under wet weather conditions, these combined sewer systems would overflow when large amounts of stormwater would enter the system. State and local authorities generally have not allowed the construction of new combined sewers since the first half of the twentieth century.
- **Separate sanitary sewers** are installed to collect wastewater only and do not provide widespread drainage for the large amounts of runoff from precipitation events. Sanitary sewers are typically built with some allowance for higher flows that occur when excess water enters the collection system during storm events. Sanitary sewers that are not watertight due to cracks, faulty seals, and/or improper connections can receive large amounts of infiltration and inflow (I/I) during wet weather. Large volumes of I/I can cause sanitary sewer overflows (SSOs) and/or operational problems at the wastewater treatment facility serving the collection system. In addition, sewage overflows can be caused by other problems such as blockages, equipment failures, broken pipes, or vandalism. Resources for overflows and peak flows at wastewater treatment plants include (1) combined sewer overflows (CSOs) and (2) sanitary sewer overflows (SSOs).

Significant increases in flow at wastewater treatment facilities caused by wet weather conditions can create operational challenges and potentially adversely affect treatment efficiency, reliability, and control of unit process operations at the treatment facility. The CSO policy encourages municipalities with combined sewers to maximize wet weather flows to the treatment plant in order to decrease uncontrolled overflows in the collection system. Resources for peak flows at wastewater treatment facilities include

- Peak flows at treatment plants–Wet weather conditions can result in peak flows to treatment plants.

Also important in the stewardship of public health by wastewater managers are approaches to

- **Integrated planning**—Voluntary opportunity for a municipality to propose to meet multiple Clean Water Act (CWA) requirements by identifying efficiencies from separate wastewater and stormwater programs and sequencing investments.

- **National pretreatment program**—The National Pretreatment Program identifies specific discharge standards and requirements that apply to sources of nondomestic wastewater discharged to a publicly owned treatment works.
- **Biosolids**—Biosolids are the nutrient-rich organic materials produced by wastewater treatment facilities.

ENVIRONMENTAL LEGISLATION

Utilities must comply with specific provisions of several environmental legislative initiatives to protect air and water, to address resource conservation and recovery matters, and to ensure the viability of the utility in emergency situations while protecting the public's right to know about a particular event. It is important to remember that while compliance has been mandated for many laws and their attendant rules, most laws/rules have resulted from public concerns about the laws' topical issues. Honest efforts in compliance and reporting requirements are key to successful implementation and build credibility within the community as the utility tells its story and explains the associated costs and programs.

The principal US laws that affect water utilities are the Clean Air Act, the Clean Water Act, the Resource Conservation and Recovery Act, the Safe Drinking Water Act, and the Emergency Planning and Community Right-to-Know Act. Other federal and state legislative acts may have articles that affect utility operations. The best information about these and other state legislative initiatives can be found through USEPA and your local regulatory agency.

Safe Drinking Water Act Framework

In the United States, the Safe Drinking Water Act (SDWA) provides an overall regulatory framework for ensuring the safety of drinking water. The US Environmental Protection Agency is responsible for the development of national drinking water regulations and the implementation of these regulations.

As it was last amended in 1996, the SDWA established a scientific, risk-based process for identifying, assessing, and managing health risks from contaminants in drinking water. Since 1975, USEPA has developed 18 major drinking water regulations addressing 91 different contaminants.

In 2010 USEPA announced a new Drinking Water Strategy that will regulate groups of contaminants and utilize several other principles to increase coordination and collaboration between USEPA and states as well as among federal statutes.

Most states are authorized (given primacy) to enforce USEPA rules when they adopt regulations no less stringent than the federal versions.

Water supply in Canada is managed at the provincial level; thus there is no national jurisdiction unless federal institutions or first nations are served. Several provincial governments have laws and regulations to address their specific requirements, including environmental issues. It is important to ensure sustainability of the water system's assets. AWWA's Canadian Affairs Council provides additional experience. In Mexico, environmental issues are addressed by both federal and state agencies. It is noted that the United States, Canada, and Mexico have many joint programs and projects that address environmental concerns of all nations in North America.

Utility managers are responsible for their organizations' compliance with provisions of all environmental laws that affect their business. Thus, it is important that a utility manager has a solid understanding as to requirements within the law and that he or she assigns authority for ensuring compliance to the appropriate subordinate area. Because the manager holds ultimate responsibility, it is imperative that he or she be kept abreast of the utility's position and needs so that funding or program priorities are adjusted to meet compliance requirements.

RISK MANAGEMENT AND INSURANCE

Each utility, as part of its basic business practices, maintains insurance to cover its property, personnel, and emergencies that might occur. Generally, this insurance coverage is maintained through an external provider that handles claims and that may provide training, voluntary inspections, or other services. Typically, utilities maintain liability, replacement, damage, and vehicle insurance through commercial carriers. Each type of insurance has some degree of regulation through state or provincial government insurance commissions, which ensure a modicum of consistency as to practices, coverage, rates, and protocols for claims. Utilities normally insure with commercial carriers that have experience in utility coverage and that assist the organization in determining the correct deductibles, ranges of benefits, and so on. A good working relationship with a carrier's agent is a strong asset for a utility manager and the risk management staff.

Some utilities prefer to self-insure, and this practice is frequently chosen for employee benefits insurance—health, life, disability, retirement, catastrophic, and so on. The utility then hires a benefits administrator to establish payment rates for services, approve procedures, and process claims. It is important that the utility manager and risk manager know the requisite state procedures for self-insured utilities and that those programs are continually monitored for fiscal accuracy and desired service levels.

Workers' compensation insurance is required of each utility, whether it is self-insured or covered under a state plan. Utility and risk managers must be highly knowledgeable about the time elements, reporting requirements, and rules and regulations of workers' compensation programs, as they differ from state to state. Workers' compensation systems are a significant resource for the utility in terms of reducing costs and identifying and correcting work procedures, equipment, or equipment usage that frequently result in on-the-job accidents and worker injury. In addition,

- Most workers' compensation agencies provide training and programs for risk managers and, in some cases, for all utility personnel. These training programs include materials and videos that address worker injury matters.
- Many agencies also provide consultation services to utilities, in many cases conducting on-site inspections or audits.
- Resources suggested by the agency that help ensure compliance with other health and safety requirements can, if implemented, result in reduced premium and payment costs.

PUBLIC COMMUNICATION AND NOTIFICATIONS

Strategic communications planning is important, especially with evolving regulatory requirements, enhanced media outreach, and growing public expectations. The following are four steps for successful communications planning:

1. Figure out what message you want to communicate.
2. Define your target audiences.
3. Assess your utility's current communications procedures.
4. Select the public relations strategies that will most effectively communicate your image and messages to your key audiences.

An effective communications plan enhances the general public's perception of the utility through frequent and focused communications with utility customers and stakeholders. Among the many benefits of a communications and customer relations plan are increased understanding and support of sustainable rate structures, greater tolerance for service interruption, better cooperation to keep construction projects on schedule, and improved response to billing issues in a timely payment of bills. AWWA's utility management standard, ANSI/AWWA G420, Communications and Customer Relations, covers the essential requirements to effectively manage communications and customer relations for a water utility.

Public Notifications

Public notification has always been a part of the Safe Drinking Water Act. USEPA requires notifications to all consumers, within 24 hours in some cases, in the event of a serious drinking water problem or emergency (e.g., contamination at levels that pose an immediate threat to human health). The clock starts ticking as soon as the water utility is notified of the violation.

AWWA Public Affairs Council (PAC) has assembled a Public Communications Toolkit containing information from public communication requirements to talking points to *Journal AWWA* articles to presentations, at http://www.awwa.org/resources-tools/public-affairs/communications-tools/public-communications-toolkit.aspx#2213177-crisis--issue-communications

The expectation of a wide range of public notifications concerning utility problems and operations has grown in recent years. As a result, many utilities are faced with many circumstances where notification is expected and required. Those circumstances can include

- Contamination
- Security breaches and plant break-ins
- Nitrates
- Security alerts (homeland security issues)
- Arsenic
- Boil-water alerts (see Figure 9-1)
- Main line breaks
- High-water alerts
- Traffic pattern changes
- Power outages

In addition, citizens are increasingly interested in knowing about utility work zone areas—what is to be done and the timetable for repairs—and weather events that may affect distribution or water quality (tornadoes, major storms, flooding, and hurricanes). Also, the public must be notified about chemical releases or spills, regardless of whether they occur in a contained area or whether they are on public property.

Maricopa County
News Release

Environmental Services Department
1001 N Central Avenue
Suite 550
Phoenix, AZ 85004
Phone 602-506-6611
Fax 602-506-1874
www.maricopa.gov/envsvc

For additional information: Johnny Diloné, MCESD Public Information Officer
602-506-6611, Cell Phone: 602-525-2423
Emily Poland, MCDPH Public Information Specialist
602-506-6607, Cell Phone: 602-722-1806

January 25, 2005

BOIL WATER ADVISORY: SAFETY MEASURES REQUIRED

Maricopa County Departments of Environmental Services and Public Health advise all water customers served by the City of Phoenix to boil water used for consumption and take necessary precautions until further notice. The Boil Water Advisory was issued this morning by City of Phoenix officials due to the detection of high sediment levels, known as turbidity, found to be entering the water supply at some City of Phoenix water treatment plants.

This precautionary health advisory is extended to all City of Phoenix water customers, including residents in Phoenix, Tolleson and other areas served by the City of Phoenix water supply.

Be advised that water needs to be at a boil for at least 5 minutes before being used in the following situations:

- Drinking water
- Washing dishes
- Brushing teeth
- Food preparation
- Making ice
- Wound care

The Boil Water Advisory affects the health and safety of all eating and drinking establishments including retail groceries operations, restaurants, schools, daycare centers, hospitals, senior care facilities, etc. Food establishments need to follow special standards to remain in operation. Please visit the following link for a list of the required items:

www.maricopa.gov/envsvc

Schools and day care centers should prevent children and staff from consuming water from water fountains. If bottled water is unavailable, necessary actions should be taken to prevent children from dehydrating, particularly after any physical activities.

-more-

Figure 9-1 A sample boil-water order issued by Maricopa County, Arizona

Source: Courtesy of Maricopa County, Arizona.

(continues)

Maricopa County NewsRelease
Add Descriptive Text Here
Page 2 of 2

"If an establishment is not capable of meeting these health and safety standards, they must remain closed until the Boiled Water Advisory is lifted by Maricopa County Environmental Services," said Al Brown, Maricopa County Environmental Services Director.

Additionally, residents affected by this current water safety situation are encouraged to follow a water conservation plan as recommended by City of Phoenix and Maricopa County officials. This may include limiting bathing or showering, washing cars, and watering lawns and plants until further notice.

High sediment or turbidity levels interfere with the disinfection process and provide a growing medium for microbial growth.

"Turbidity may indicate the presence of disease causing organisms, but it is important to know that water may still look fairly normal," Brown said. "Therefore it is essential that residents follow proper water and sanitation precautions, until notified by County Environmental Health officials that the safety of the water has been assured."

"This has the potential to be a serious public health issue, but we're taking all necessary precautions to prevent disease in the county," said Dr. Doug Campos-Outcalt, Medical Director for the Maricopa County Department of Public Health. "The role of the Department of Public Health in such situations is to advise residents of potential threats to their health, to advise them of appropriate precautions to take to protect their health and safety, and to communicate to them when the health threat is over."

The MCDPH is working with the community and is notifying health care providers and facilities to watch for patients with gastrointestinal disease symptoms. Residents who may experience nausea, vomiting, diarrhea, and abdominal pain should consult a health care provider as disease symptoms could occur from 1 to 12 days after disease exposure.

Please be aware that this Boil Water Advisory will be in effect until lifted by the Maricopa County Environmental Services Department.

For more information private citizens may call the City of Phoenix Water Department customer service line: (602) 262-6251.

The Arizona Department of Environmental Quality has the following link for emergency disinfection of drinking water: *http://water.azdeq.gov/envion/water/dw/download/desinfec.pdf*

###

Figure 9-1 A sample boil-water order issued by Maricopa County, Arizona (continued)

Case Study: *Salmonella* Outbreak in Alamosa, Colorado, Water Supply

An outbreak of waterborne disease associated with *Salmonella* in drinking water struck Alamosa, Colorado, during March and April 2008. The city of Alamosa's public water system that supplies drinking water to the community became contaminated with *Salmonella* bacteria. Alamosa's population is about 8,900. The outbreak resulted in 442 reported illnesses, 122 of which were laboratory-confirmed, and 1 death. Epidemiological estimates suggested that up to 1,300 people may have been ill. Alamosa's drinking water comes from deep artesian wells in an aquifer considered to be a protected groundwater source. Prior to the outbreak, the city's drinking water was not chlorinated for disinfection. A statewide response to the outbreak lasting about one month involving numerous responders was coordinated using the National Incident Management System (NIMS) and Incident Command System (ICS). During the outbreak residents were advised to drink bottled water, and then the entire water system was flushed and disinfected with chlorine to kill the *Salmonella* bacteria. A boil-water order followed the bottled water order, and it remained in place until tests confirmed that the city's water was safe to drink again. Because of the quick operational response and the changes made to the physical infrastructure of the Alamosa public water system, the water in Alamosa has been safe to drink since April 2008. The city has since installed advanced treatment processes and improved system operations. (Falco and Williams 2009)

When an event occurs that requires public notification, the utility is usually busy exercising its technical capacity to respond to the issue, leaving little or no time to develop a written document describing the situation. It is in the manager's best interest to develop a template that contains the required language and leaves blank spaces where the critical, incident-specific information is entered. This template approach facilitates preparing the notifications as an administrative process rather than having to assemble technical or subject-matter experts to draft the language when their expertise is sorely needed elsewhere. The language that is necessary for the template—in addition to that which is required—includes what the notification is about, who is affected, what the water supply can be used for (e.g., sanitation), and where additional information can be obtained (contact names and telephone numbers). After notification is given, concerned citizens are likely to call the utility, and customer service staff must have essential information about the notification in order to respond professionally to inquiries.

When public notification is required, utility managers are likely to be in contact with the media, including radio, television, and newspapers, but there are other tools available to fulfill notification requirements. The utility's web site can carry a copy of the notification, and by having it prepared, utility personnel can quickly duplicate flyers for door-to-door distribution to customers or for public posting. Other contact information should include the utility's emergency management team, the health department, police and fire agencies (city, county, and state or provincial), and other outside agencies that might assist in the utility's response or require special notification, such as the Red Cross or the Department of Transportation.

Consumer Confidence Reports (CCRs)

The 1996 amendments to the Safe Drinking Water Act contained requirements to ensure consumers are informed about their drinking water. The law requires utilities to mail a Consumer Confidence Report (also known as a *water quality report*) to their customers every year. Both the SDWA and the enabling Consumer Confidence Report (CCR) Rule of 1998 list specific information that must be included in the report.

Based on a review of the CCR Rule, the US Environmental Protection Agency recently issued an interpretative memo providing for the electronic delivery of CCRs (eCCR). There are many important aspects of eCCR delivery that all utilities should be aware of. In support of electronic delivery of CCRs and to help utilities research delivery options that meet USEPA requirements, AWWA has developed a complementary *Electronic CCR Best Practices Guide* as a resource to utilities interested in developing eCCR programs.

A utility manager has an opportunity to use the CCR as a vehicle to communicate with stakeholders about utility activities, improvements in water quality and service, and any other information that would be of interest to them.

EMERGENCY PREPAREDNESS PLANNING

Preparedness is defined by the Department of Homeland Security (DHS) and the Federal Emergency Management Agency (FEMA) as "a continuous cycle of planning, organizing, training, equipping, exercising, evaluating, and taking corrective action in an effort to ensure effective coordination during incident response." Utility managers have to be prepared for the worst. You cannot control the threat, but you can manage the response. The key to managing risks to water supply and utility operations is to make sure that responses are developed to match threats as they evolve.

After the September 11 terrorist attacks, drinking water systems were mandated to complete vulnerability assessments (VAs) and emergency response plans under provisions of the Public Health Security and Bioterrorism Preparedness Act of 2002. These VAs take an all-hazards approach to risk assessment and help the utility make a more realistic investment in security. The methodology in ANSI/AWWA J100, Risk and Resilience Management for Water and Wastewater Utilities, applies the all-hazards framework, recognizing that terrorism may not be the highest-probability threat in most locales relative to the known probabilities of other incidents such as hurricanes, earthquakes, or floods. The risk management strategies implemented for responding to natural hazards often have a direct value in preparing a utility for actions that may be necessary when responding to a terrorism incident.

Denver Water's Experience

What does not get valued does not get managed, which is precisely what motivated Denver Water to integrate security into its risk management program. The utility came to realize that the department owned real property in multiple counties, including water rights and treatment operations and equipment that were valued in excess of $6.5 billion. These critical assets represent a significant value that is worth protecting from loss. The approach and justification applied by Denver Water did not require a knock on the door by Mother Nature or some malevolent actor. It is rooted in leadership making a commitment to manage the risk exposures that could compromise the operational integrity of the system. (*Source: Journal AWWA* 2014)

Utility emergency response plans should include policy, protocol, and procedures and are required to be in writing and accessible and understandable by all employees. When public notification is required, utility staff cannot be mired in locating the adopted procedures and protocols. They need to be dealing with the situation. This part of the process is the responsibility of managers and supervisors, who must ensure that all staff members know their responsibilities and know that they will be held accountable for complying with those requirements.

One of the most important tasks of a utility manager is to conduct practice drills of protocols before an actual emergency occurs. Employees should be trained on utility procedures, governing policies and protocols, and other areas as appropriate to ensure the desired response level. Training topics include personal protective equipment, handling difficult people, working under stressful conditions, workplace violence, and the utility's basic commitment to quality service.

In serious emergencies, managers must ensure that they have access to contractors, vendors, consultants, laboratories, equipment rental agencies, and emergency equipment contractors for unique requirements. The requirements include snow or debris removal, environmental cleanup, health and wellness follow-up, and special equipment, such as cranes, derricks, large haulers, and related equipment (e.g., water-haulers).

OCCUPATIONAL HEALTH AND SAFETY

Utility managers must strive to operate their organizations in compliance with environmental, health, and safety laws and regulations. More important, managers are responsible for ensuring a safe, healthy, and productive workplace through training, compliance programs, and management systems that identify and address risk factors. Employees are a utility's most valuable resource, and their ability to work safely and in a safe environment is the main reason for workplace safety programs. Safeguarding the general public from construction sites and operations is another reason for the programs. Managers play a key role in

- Developing and administering industrial hygiene and safety
- Defining compliance requirements for each of their operations
- Supplying guidance to supervisors and employees in implementing the adopted health and safety program
- Regularly measuring their organization's progress, making corrections to improve conditions or processes, and ensuring that staff compliance with health and safety protocols is a measurable part of performance management

AWWA Safety Policy

American Water Works Association (AWWA) considers a safe work environment to be vital for everyone who works in the water industry, including those responsible for the maintenance and protection of our public water supplies. Employers must fully adopt an effective health and safety program developed according to established risk factors and regulatory compliance, and the program must be integrated and maintained throughout the workplace. It is everyone's duty to comply, uphold and continually improve all safety and health regulations, standards and procedures.

AWWA believes that effective health and safety programs successfully balance human, environmental, cultural and political challenges in achieving health and safety excellence in all aspects of the water industry and is committed to providing educational and support materials to members to further this aim.

Adopted by the Board of Directors Feb. 2, 1997, and revised Jan. 18, 2004, and Jan. 17, 2010. Revised June 7, 2015.

A formal health, safety, and environment (HSE) program is a reliable and effective means of consistently protecting workers from occupational injuries and illnesses. Formal written programs can address specific hazards and prevention measures (e.g., fall prevention) as well as specific regulatory mandates that require a written program (e.g., hazard communications, respiratory protection). Furthermore, there is widespread agreement that written management systems improve organizational performance, including performance in occupational health and safety. AWWA Manual M3, *Safety Management for Utilities,* is a complete resource for establishing written HSE management systems.

REFERENCES

AWWA. 2006. *Asset Management Planning and Reporting Options for Water Utilities.* Denver, CO: AWWA.

American Water Works Association (AWWA). 2014. Manual M3. *Safety Management for Utilities.* Denver, CO: AWWA.

Falco R., and S.I. Williams. 2009. Waterborne Salmonella Outbreak in Alamosa, Colorado: March and April 2008—Outbreak Identification Response and Investigation." Denver, CO: Colorado Department of Public Health and Environment, Safe Drinking Water Program, Water Quality Control Division.

USEPA. 2017. *Effective Utility Management (EUM).* Washington, DC: USEPA.

US General Accountability Office (GAO). 2004. Water Infrastructure: Comprehensive Asset Management Has Potential to Help Utilities Better Identify Needs and Plan Future Investments. GAO-04-461. Washington, DC: US GAO.

This page intentionally blank.

Appendix **A**

Information Technology Security: K.M. Morley's Perspective

Information technology security, known also as *cybersecurity* or *IT security*, is the protection of information systems from theft or damage to the hardware, the software, and to the information on them, as well as from disruption or misdirection of the services they provide. The following *Journal AWWA* article by Morley addresses the water–wastewater response to the need for security and preparedness, including controlling physical access to the hardware, as well as protecting against harm that may come via network access, via data and code injection, and by operators.

Emergency Management: You Can't Control the Threat, But You Sure Can Manage the Response

Kevin M. Morley, *Journal WWA*, August 2014 (Abridged)

The key to managing risks to water supply and utility operations is to make sure that responses are developed to match threats as they evolve.

The water sector is no stranger to risk and resilience management; it is the central principle of our mission as stewards of public health and the environment. Managing the societal risks associated with waterborne disease were the underpinnings of massive transformations in public health.

RISK GOVERNANCE

Effective risk management requires leadership and a willingness to take ownership of the full risk portfolio. Water utilities operate in an all-hazards environment and do not have the option of selectively managing risk. This is a central tenet of the ANSI/AWWA G430, Security Practices for Operations and Management (2009), and ANSI/AWWA G440, Emergency Preparedness Practices (2011), which require clearly defined commitments from management through policy and budget. What does not get valued does not get managed, which is precisely what motivated Denver Water to integrate security into its risk management program (Jensen, 1982). The utility came to realize that the department owned real property in multiple counties, including water rights, and treatment operations and equipment that were valued in 1975 at approximately $1.5 billion (~ $6.4 billion in 2013 dollars). These critical assets represent a significant value that is worth protecting from a loss. The approach and justification applied by Denver Water did not require a knock on the door by Mother Nature or some malevolent actor. It is rooted in leadership making a commitment to manage the risk exposures that could compromise the operational integrity of the system. This simple example demonstrates the nexus between security preparedness and asset management, all of which represent elements of total risk management (Liner et al., 2012; McLaughlin, 2011).

The convergence of these two seemingly different business practices has been facilitated by the need for a more complete assessment of risk and resilience. After the September 11 terrorist attacks, drinking water systems were mandated to complete vulnerability assessments (VAs) and emergency response plans under provisions of the Public Health Security and Bioterrorism Preparedness Act of 2002. These early VAs were not complete risk assessments as they were solely focused on potential terrorist threats. The implication was a disproportionate investment in physical security measures relative to steps that could otherwise prepare the utility for other high-probability threats such as floods, hurricanes, and earthquakes. In recognition of this limitation, the methodology in ANSI/AWWA J100, Risk and Resilience Management for Water and Wastewater Systems (2010), applies an all-hazards framework (Morley, 2010). The first phase of this assessment identifies which assets are operationally critical and subsequently identifies the loss value. This is essentially the same information captured and documented in a robust asset management program. The net result of a J100 assessment is the identification of measures that generate the highest net benefit to reduce risk and/or enhance resilience. This is accomplished by requiring a worst-reasonable-case analysis of a broad class of threats. In calculating risk, the J100 assessment applied the standard calculation (Eq 1; DHS, 2008) and represents the result in a cost–benefit format to facilitate utility decision-making.

$$\text{Risk (R)} = f\ [\text{Threat (T), Vulnerability (V), Consequence (C)}] \quad \text{(Eq 1)}$$

The J100 all-hazards risk management framework recognizes that terrorism may not be the highest-probability threat in most localities relative to the known probabilities of other incidents such as hurricanes, earthquakes, tornadoes, or floods. In fact, the risk management strategies implemented for responding to natural hazards often have direct value in preparing a utility for actions that may be necessary when responding to a terrorism incident. Water systems, including Birmingham Water Works Board (Starkey et al., 2011), City of Richmond, Va. (Warren et al., 2012), Chicago Department of Water Management (Chiczewski and Gayle, 2013), Long Beach Water Department, American Water (Binning, 2013), DC Water (Reeves, 2013), and Minneapolis (Rezania et al., 2014), have all undertaken risk-and-resilience assessments based on the ANSI/AWWA J100-10 standard as a means of identifying high-risk threats and prioritizing asset management decisions to optimize risk and resilience measures.

In most instances water utility owners/operators are not in a position to influence the intent of a terrorist or other malevolent actor, not to mention the path of a tornado. Yet an inordinate amount of attention and resources is being allocated to the threat du jour rather than focusing attention on measures to prepare for and respond to potential consequences independent of cause. There is a need to be prepared to manage the effect of any incident, which necessitates a need to better understand an entity's level of resilience. The Boy Scouts of America motto—Be Prepared—is the simplest articulation of this principle. Resilience can also be reframed as an assessment of an entity's lack of preparedness.

By incorporating resilience into the risk management framework, a utility can improve its response and recovery strategies, thereby mitigating the potential for loss of service. This is the basis for the development of the Utility Resilience Index (URI), which provides a simple, high-level analysis of resilience in a manner that is readily accessible to utility decision-makers (Morley, 2012a). The URI examines key "deficiencies in preparedness," a dimension of risk critical in supporting an all-hazards management strategy. The 12 indicators that make up the URI are based on operational and financial factors that either directly or indirectly influence continuity of operations (Table A-1).

Application of the URI has revealed strengths and opportunities in utility actions to enhance capacity to respond to and/or cope with the effects of an incident independent of the cause. Table A-2 provides some insight on the resilience from a cross section of utilities of all sizes and geographical distribution. These data validate the observation from multiple events in which power is often the biggest rate-limiting factor in water utility recovery efforts (AWWA, 2013; Morley, 2012b). Recent events, be it drought or contamination, also indicate a need to fully assess our sector capabilities in providing emergency water supply (CDC and AWWA, 2012; USEPA, 2011; Romer et al., 1965).

MOVING FORWARD

Maintaining vigilance and being adaptive to new threats are essential to risk management. The most challenging threat facing all critical infrastructure sectors is cyber-based because it does not require the perpetrator to be physically present and because of increased automation. The latter means a huge increase in potential points of entry spanning every business unit, including email, accounting and business systems, supervisory control and data acquisition, mobile devices, and motor controls. The nature of water-sector operations as a lifeline that is essential to daily life means that this multidimensional threat must not be taken lightly (Dakin et al., 2009). Actions have been taken to enhance the water sector's management of this risk through guidance that provides a means to grapple with this complex issue (AWWA, 2014; Roberson and Morley, 2014).

Ensuring the integrity of the water supply remains a constant, including preparedness for anomalous events that threaten basic operations but also public confidence. The fundamental action is to mitigate threats of contamination, malevolent or accidental in nature. This need is underscored by guidance and resources developed by the US Environmental Protection Agency (USEPA, 2014) to detect, model, and respond to intentional contamination. Despite all of the work that has been done, many challenges remain in developing effective strategies and solutions for response and recovery (Sullivan and Schmelling, 2012; CIPAC, 2008). Incidents such as the Elk River chemical spill in West Virginia have highlighted the need for constant vigilance and realigning the scope of source water–protection planning efforts (Digiano and Grayman, 2014; Sham et al., 2010). The event also exposed limitations in the regulatory matrix designed to protect water systems because of variable oversight for some potentially hazardous materials, utility access to information regarding materials that could harm supplies is not efficient, and utility notification following an incident affecting water supplies is inconsistent. This requires a collective strategy that leverages the information accessible to first responders to advise of potential threats to preparedness and mitigation actions that may be deployed by a utility and/or state/federal responders.

Benjamin Franklin once said that by "failing to prepare, you are preparing to fail." This subtle statement emphasizes what is likely the most essential of all security and preparedness actions in which a utility can invest: personnel. Utility staff must have clearly defined roles and responsibilities and have received the appropriate training. This seems like a simple task, but all too often plans are developed but rarely exercised. The value of a plan is the act of planning itself, in which utility staff examine and familiarize themselves with the response protocols. The after-action report addressing Massachusetts Water Resources Authority's response to a massive 10-ft-diameter coupling failure makes the value of training very clear when staff reported that because of the training received, the event felt more like a drill (AMWA et al., 2011). Staff members had a working knowledge of expectations that mitigated the excitement and chaos factors that can negatively compound during a response when staff is not prepared. This principle is also emphasized in ANSI/AWWA G440, Emergency Preparedness Practices. Developing experience with emergency response protocols has also been a factor in enhancing the effectiveness of mutual aid responses and related incidents that require quick, decisive action by utility staff and partners (Casson et al., 2011; Whitler and Stormont, 2011).

All in all, there eventually will be a need to react, respond, and recover in a manner that sustains the communities we serve as water utilities as well as the personnel we employ and their families. Just like at home, this commitment must begin at the top with management—aka head of household. If utility managers do not buy into a culture of preparedness, their organizations will not invest wisely in the risk-and-resilience management strategies necessary to mitigate future loss to the utilities and the communities they serve.

Table A-1 Indicators for tabulating URI

Indicator Description
O_1: Emergency response plan
O_2: National incident management system
O_3: Mutual aid and assistance
O_4: Emergency power
O_5: Minimum daily demand
O_6: Critical parts and equipment
O_7: Critical staff
F_1: Business continuity plan
F_2: Bond rating
F_3: GASB 34 condition assessment
F_4: Unemployment
F_5: Median household income

Notes: GASB—Governmental Accounting Standards Board; URI—Utility Resilience Index.

Table A-2 URI data for emergency power and minimum daily demand

URI Indicator	Distribution of Maximum Reported Value ($n = 42$)—%				
	None	Up to 24 hr	> 24–48 hr	> 48–72 hr	> 72 hr
O_4: Emergency power	12	5	7	36	40
O_5: Minimum daily demand	2	26	26	24	21

Source: Morley, 2012a.
URI—Utility Resilience Index.

ABOUT THE AUTHOR

Kevin M. Morley is the security and preparedness program manager at AWWA and is based in Washington, D.C.; kmorley@awwa.org.

REFERENCES

AMWA (Association of Metropolitan Water Agencies), Water Research Foundation, and WaterISAC, 2011. Multi-agency Response to a Major Water Pipe Break: A Massachusetts Case Study and Evaluation. www.amwa.net/sites/default/files/MWRA_Case_Study_7-7-11.pdf (accessed May 2014).

AWWA, 2014. *Process Control System Security Guidance for the Water Sector.* AWWA, Denver, CO.

AWWA, 2013. *WARN: Sandy After-Action Report.* AWWA, Denver, CO.

AWWA, 2011. ANSI/AWWA G440-11: Emergency Preparedness Practices. AWWA, Denver.

AWWA, 2010. ANSI/AWWA J100-10: Risk Analysis and Management for Critical Asset Protection (RAMCAP®) Standard for Risk and Resilience Management of Water and Wastewater Systems. AWWA, Denver.

AWWA, 2009. ANSI/AWWA G430-09: Security Practices for Operations and Management. AWWA, Denver.

Binning, D., 2013. Wastewater Resilience: Measurement. *Proc. 2013 WEFTEC, Chicago, Ill.*

Casson, L.W., Morley, K.M., States, S.J., Watson, J.W., and Roberson, J.A., 2011. Increasing Preparedness for Water Utilities, Local Health Departments, and First Responders. *Journal of Emergency Management,* 9:5:39. http://dx.doi.org/10.5055/jem.2011.0072.

CDC (Centers for Disease Control and Prevention) and AWWA, 2012. *Emergency Water Supply Planning Guide for Hospitals and Health Care Facilities.* US Department of Health and Human Services, Atlanta.

Chiczewski, R., and Gayle, P., 2013. Lessons Learned Conducting a J-100 RAMCAP Threat and Vulnerability Assessment for the Chicago Department of Water Management. *Proc. AWWA 2013 Distribution Systems Symposium and Security and Emergency Preparedness Conference and Exposition, Itasca, Ill.*

CIPAC (Critical Infrastructure Partnership Advisory Council), 2008. Water Sector Decontamination Priorities: Recommendations and Proposed Strategic Plan of the Water Sector Decontamination Working Group. www.awwa.org/portals/0/files/legreg/security/decontaminate.pdf (accessed May 2014).

Dakin, R., Newman, R., and Groves, D., 2009. The Case for Cyber Security in the Water Sector. *Jour. AWWA*, 101:12:30.

Dappert, A.F., 1942. New York State Mutual Aid Plan for Water Service in Case of Emergencies. *Jour. AWWA*, 34:2:189.

Devendorf, E., 1947. Mutual Aid Program for Peacetime [N.Y.]. *Jour. AWWA*, 39:6:520.

Devendorf, E., 1943. Mutual Aid in New York in 1942. *Jour. AWWA*, 35:2:201.

DHS (Department of Homeland Security), 2008. Water Sector Specific Plan in Support of the National Infrastructure Protection Plan. www.dhs.gov/xlibrary/assets/nipp-ssp-water-2010.pdf (accessed May 2014).

Digiano, F., and Grayman, W.M., 2014. Can We Better Protect Vulnerable Water Supplies? *Jour. AWWA*, 106:4:28. http://dx.doi.org/10.5942/jawwa.2014.106.0067.

Elwell, F., Cotruvo, J.A., Kelley, W.A., LaFrankie, J.V., and Miller, W.H., 1984. Is Vandalism a Threat to Water Utilities? *Jour. AWWA*, 76:7:19.

Geupel, L.A., 1943. The Mutual Aid Plan in Indiana. *Jour. AWWA*, 35:2:227.

Geupel, L.A., 1941. Utility Co-ordination in Emergencies. *Jour. AWWA*, 33:6:1079.

Hoover, J.E., 1941. Water Supply Facilities and National Defense. *Jour. AWWA*, 33:11:1861.

Jensen, R.G., 1982. Creating the Denver Water Department's Security Program. *Jour. AWWA*, 74:10:508.

King, S., 1983. *Different Seasons*. New American Library, New York.

Liner, B., de Monsabert, S., and Morley, K., 2012. Strengthening Social Metrics Within the Triple Bottom Line of Sustainable Water Resources. *World Review of Science, Technology and Sustainable Development*, 9:1:74.

McGuire, M.J., 2013. *The Chlorine Revolution: Water Disinfection and the Fight to Save Lives.* AWWA, Denver, CO.

McLaughlin, J.W., 2011. Security and Asset Management: A Daily Double to Eliminate Jeopardy. *Jour. AWWA*, 103:11:28.

Morley, K.M., 2013. Water Sector Security and Preparedness. *Jour. AWWA*, 105:3:38.

Morley, K.M., 2012a. Evaluating Resilience in the Water Sector: Application of the Utility Resilience Index (URI). Doctoral dissertation. ProQuest Dissertations and Theses database (UMI No. 3506045).

Morley, K.M., 2012b. A Lesson in Resilience From Derecho. *Jour. AWWA*, 104:9:20.

Morley, K.M., 2010. Advancing the Culture of Security and Preparedness in the Water Sector. *Jour. AWWA*, 102:6:34.

Morley, K.M., 2009. An Evolving Culture of Security and Preparedness in the Water Sector. *Jour. AWWA*, 101:1:32.

Morley, K.M., 2007. A Vision and Direction for Water Sector Security and Preparedness Initiative. *Jour. AWWA*, 99:2:38.

Morley, K., and Riordan, R., 2006. Utilities Helping Utilities: An Action Plan for Mutual Aid and Assistance Networks for Water and Wastewater Utilities. AWWA, Denver. www.NationalWARN.org (accessed May 2014).

Moyer, J., and Novick, K., 2012. Introducing a New Resource for Water and Wastewater System Business Continuity Planning. *Jour. AWWA*, 104:3:37.

Ownbey, P.J., Schaumberg, F.D., and Klingeman, P.C., 1988. Ensuring the Security of Public Water Supplies. *Jour. AWWA*, 80:2:30.

Peisch, K., 1977. Water-quality Problems From Vandalism at Storage Reservoirs. *Jour. AWWA*, 69:6:309.

Potter, H.H., Devendorf, E., and Murdoch, J.H. Jr., 1942. The Mutual Aid Plan at Work. *Jour. AWWA*, 34:8:1173.

Reeves, J., 2013, Wastewater Resilience. *Proc. 2013 WEFTEC, Chicago, Ill.*

Rezania, S., Bankston, A., and Spence, S., 2014. Eating the Risk Assessment Process One Bite at a Time. AWWA 2014 Annual Conference and Exposition, Boston.

Riordan, R.A., 1995. Mutual Aid and Emergency Response for Water Utilities. *Jour. AWWA*, 87:5:52.

Roberson, J.A., and Morley, K.M., 2014. A Simple Action Plan for Utilities to Secure Their Process Control Systems. *Jour. AWWA*, 106:2:23.

Roberson, J.A. and Morley, K.M., 2006. Water Security: Shifting to an All-hazards Resiliency Approach. *Jour. AWWA*, 98:5:46.

Romer, H., De Zuane, J., and Schaible, G.J. Jr., 1965. Emergency Water Supplies for New York City. *Jour. AWWA*, 57:9:1099.

Sham, C.H., Gullick, R.W., Long, S.C., and Kenel, P.P., 2010. *Operational Guide to AWWA Standard G300, Source Water Protection*. AWWA, Denver, CO.

Starkey, S., and Rhaly, J., 2011. Assessing Risk and Resilience. *Proc. AWWA 2011 Distribution System Symposium and Water Security Conference, Nashville, Tenn.*

States, S., 2010. *Security and Emergency Planning for Water and Wastewater Utilities.* AWWA, Denver, CO.

Sullivan, J.P., and Schmelling, D., 2012. Report From the Contamination Warning System CIPAC Workgroup. *Jour. AWWA*, 104:10:30.

USEPA (US Environmental Protection Agency), 2014. *Water Security Initiative*. http://water.epa.gov/infrastructure/watersecurity/lawsregs/initiative.cfm (accessed May 2014).

USEPA, 2011. *Planning for an Emergency Drinking Water Supply* (EPA 600/R-11/054). www.awwa.org/Portals/0/files/resources/water%20knowledge/rc%20emergency%20prep/Emergencywater.PDF (accessed May 2014).

Warren, L.P., Steidel, R.C., Black, S., and Hatchett, J., 2012, Experiences of a Large City in Conducting a Water Vulnerability Assessment Using the J-100 RAMCAP Methodology. *Proc. AWWA 2012 Distribution Systems Symposium and Water Security and Emergency Preparedness Conference and Exposition, St. Louis, Mo.*

Whitler, J., and Stormont, C., 2011. Lessons Learned from WARN Tabletop Exercises. *Jour. AWWA*, 103:12: 24.

Index

D

E

F

G

H

I

AWWA Manuals

M1, *Water Rates, Fees, and Charges*, #30001

M2, *Instrumentation and Control*, #30002

M3, *Safety Management for Water Utilities*, #30003

M4, *Water Fluoridation Principles and Practices*, #30004

M5, *Water Utility Management*, #30005

M6, *Water Meters—Selection, Installation, Testing, and Maintenance*, #30006

M7, *Problem Organisms in Water: Identification and Treatment*, #30007

M9, *Concrete Pressure Pipe*, #30009

M11, *Steel Pipe—Design and Installation*, #30011

M12, *Simplified Procedures for Water Examination*, #30012

M14, *Backflow Prevention and Cross-Connection Control: Recommended Practices*, #30014

M17, *Installation, Field Testing, and Maintenance of Fire Hydrants*, #30017

M19, *Emergency Planning for Water Utilities*, #30019

M20, *Water Chlorination/Chloramination Practices and Principles*, #30020

M21, *Groundwater*, #30021

M22, *Sizing Water Service Lines and Meters*, #30022

M23, *PVC Pipe—Design and Installation*, #30023

M24, *Planning for the Distribution of Reclaimed Water*, #30024

M25, *Flexible-Membrane Covers and Linings for Potable-Water Reservoirs*, #30025

M27, *External Corrosion Control for Infrastructure Sustainability*, #30027

M28, *Rehabilitation of Water Mains*, #30028

M29, *Water Utility Capital Financing*, #30029

M30, *Precoat Filtration*, #30030

M31, *Distribution System Requirements for Fire Protection*, #30031

M32, *Computer Modeling of Water Distribution Systems*, #30032

M33, *Flowmeters in Water Supply*, #30033

M36, *Water Audits and Loss Control Programs*, #30036

M37, *Operational Control of Coagulation and Filtration Processes*, #30037

M38, *Electrodialysis and Electrodialysis Reversal*, #30038

M41, *Ductile-Iron Pipe and Fittings*, #30041

M42, *Steel Water-Storage Tanks*, #30042

M44, *Distribution Valves: Selection, Installation, Field Testing, and Maintenance*, #30044

M45, *Fiberglass Pipe Design*, #30045

M46, *Reverse Osmosis and Nanofiltration*, #30046

M47, *Capital Project Delivery*, #30047

M48, *Waterborne Pathogens*, #30048

M49, *Quarter-Turn Valves: Torque, Head Loss, and Cavitation Analysis*, #30049

M50, *Water Resources Planning*, #30050

M51, *Air-Release, Air/Vacuum, and Combination Air Valves*, #30051

M52, *Water Conservation Programs—A Planning Manual*, #30052

M53, *Microfiltration and Ultrafiltration Membranes for Drinking Water*, #30053

M54, *Developing Rates for Small Systems*, #30054

M55, *PE Pipe—Design and Installation*, #30055

M56, *Nitrification Prevention and Control in Drinking Water*, #30056

M57, *Algae: Source to Treatment*, #30057

M58, *Internal Corrosion Control in Water Distribution Systems*, #30058

M60, *Drought Preparedness and Response*, #30060

M61, *Desalination of Seawater*, #30061

M63, *Aquifer Storage and Recovery*, #30063

M65, *On-Site Generation of Hypochlorite*, #30065

M66, *Cylinder and Vane Actuators and Controls—Design and Installation*, #30066

M68, *Water Quality in Distribution Systems*, #30068